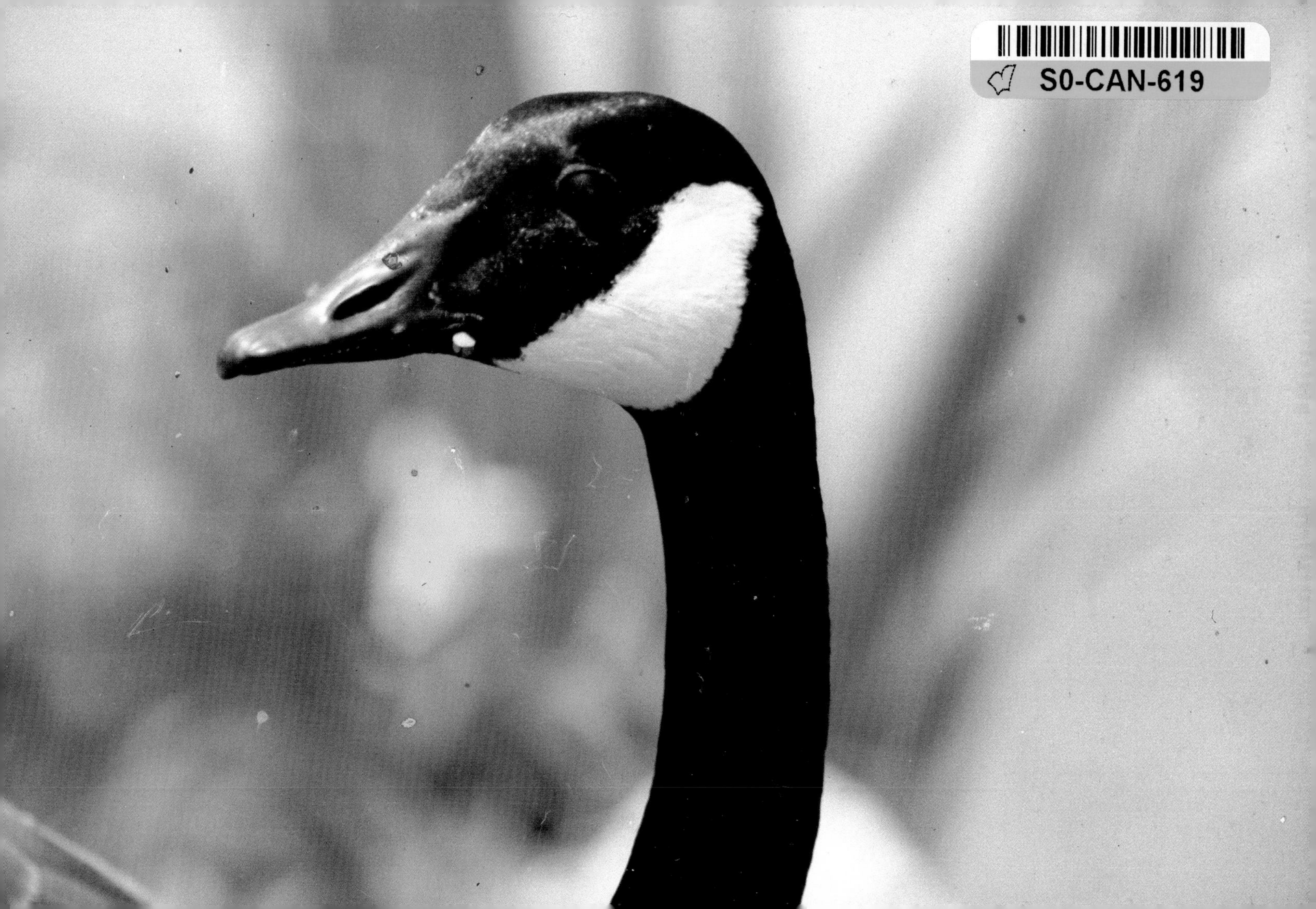

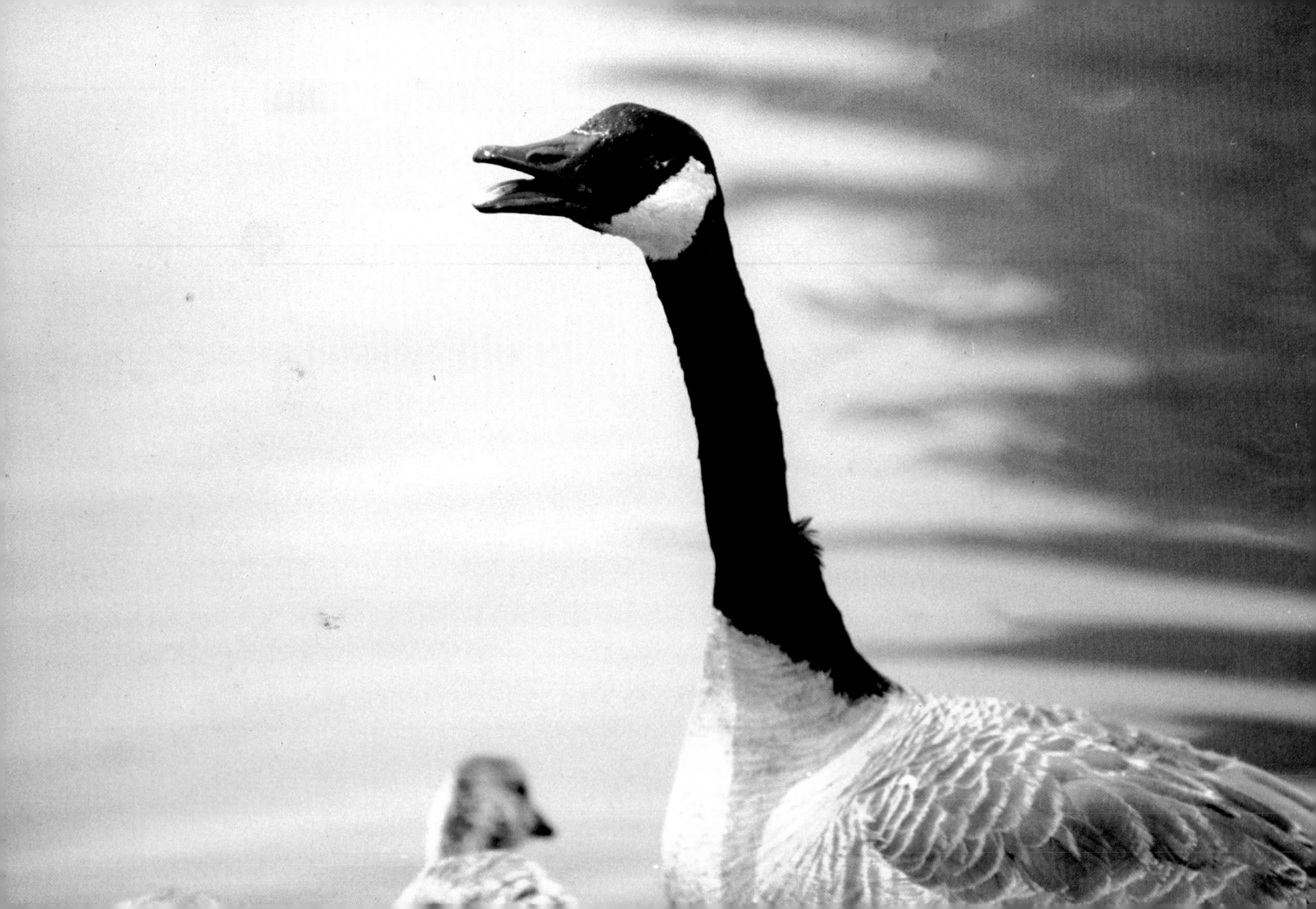

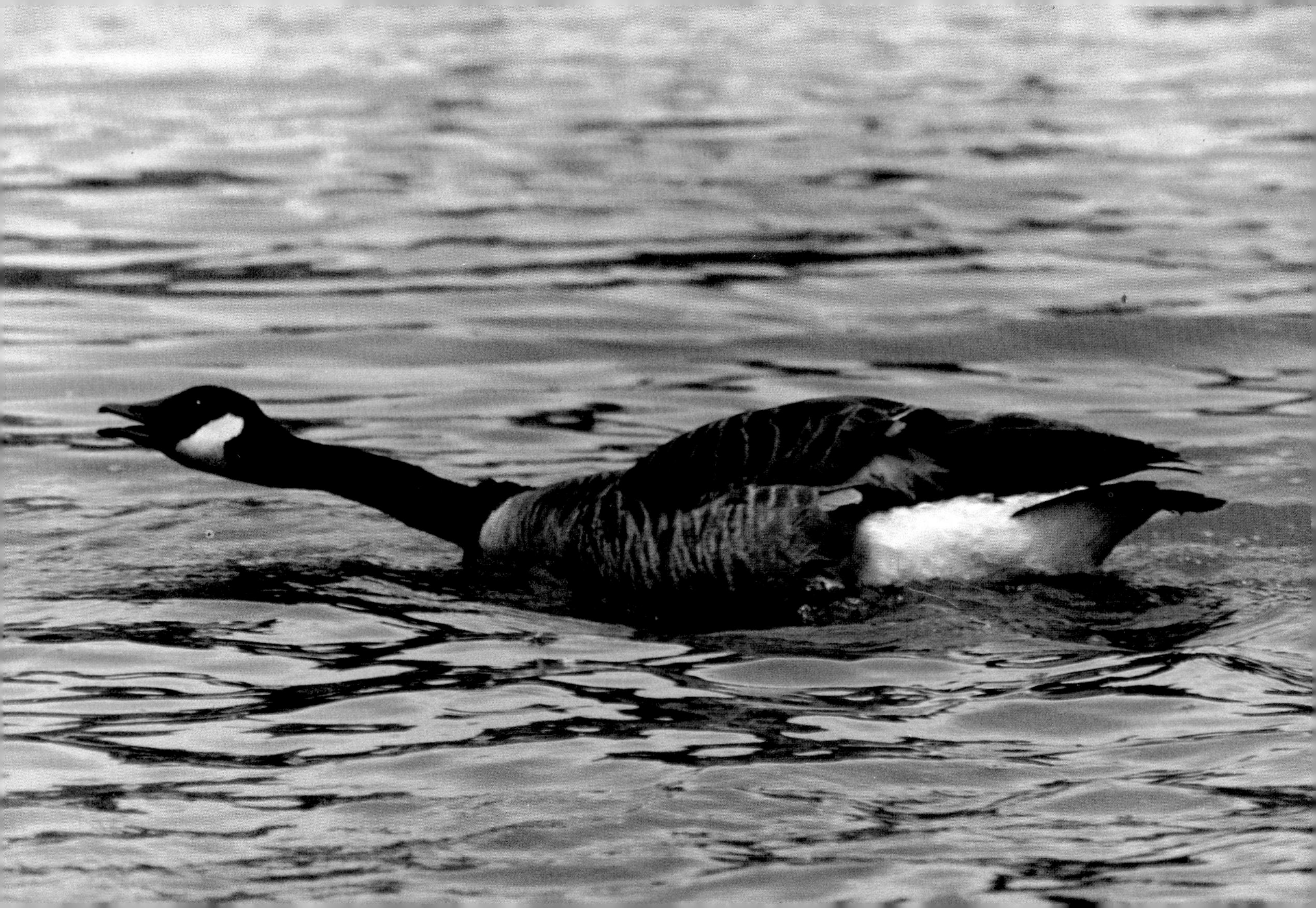

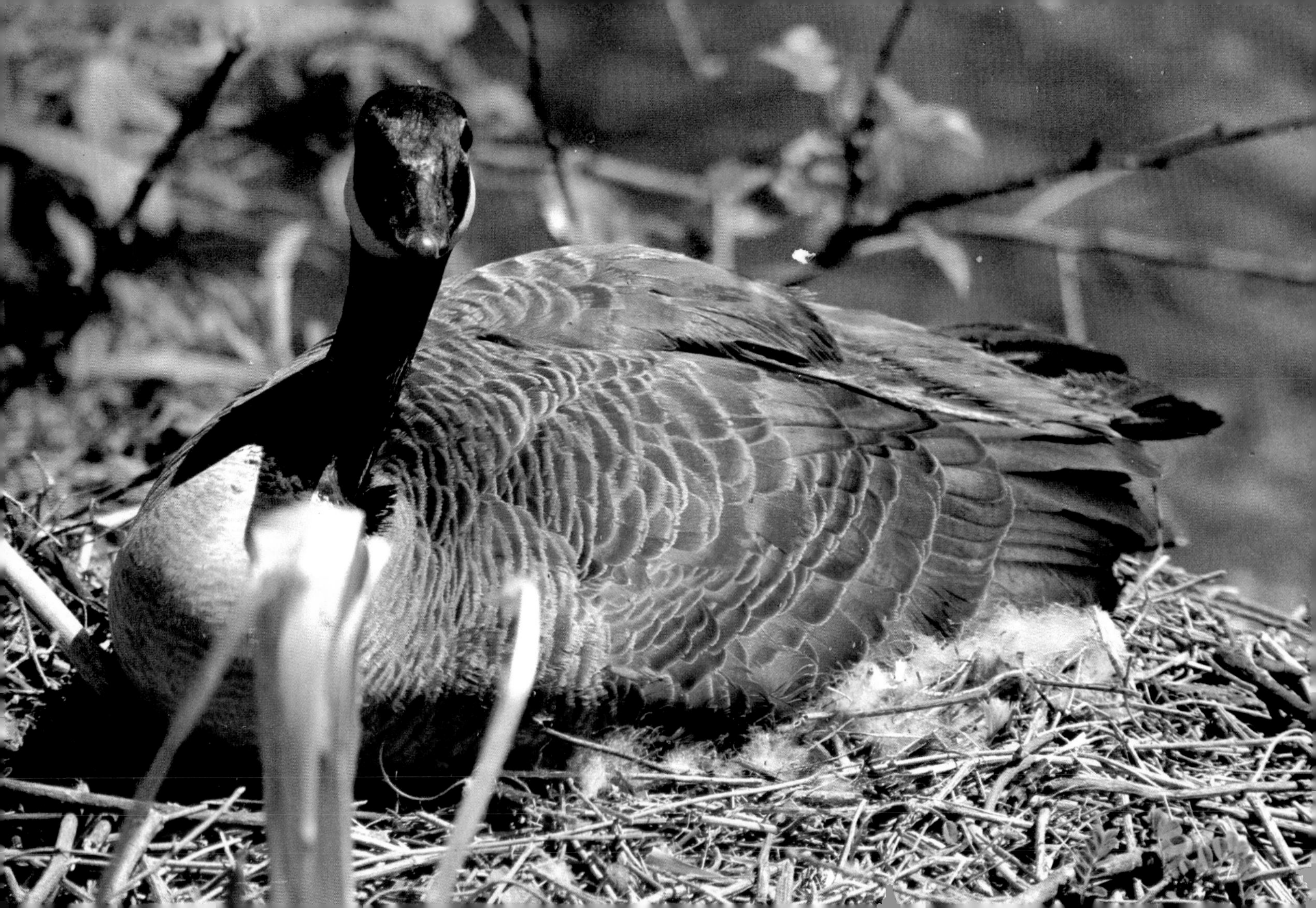

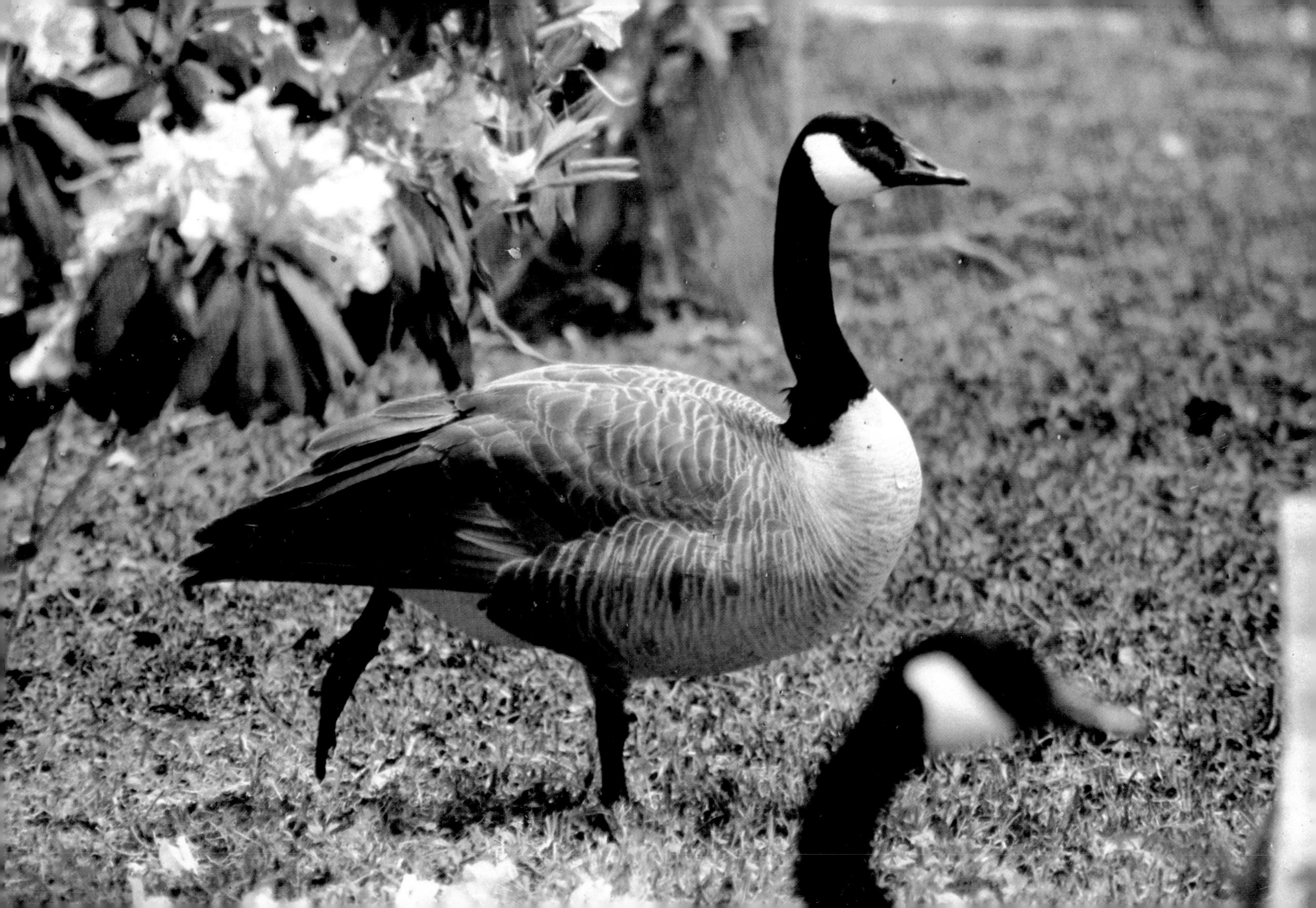

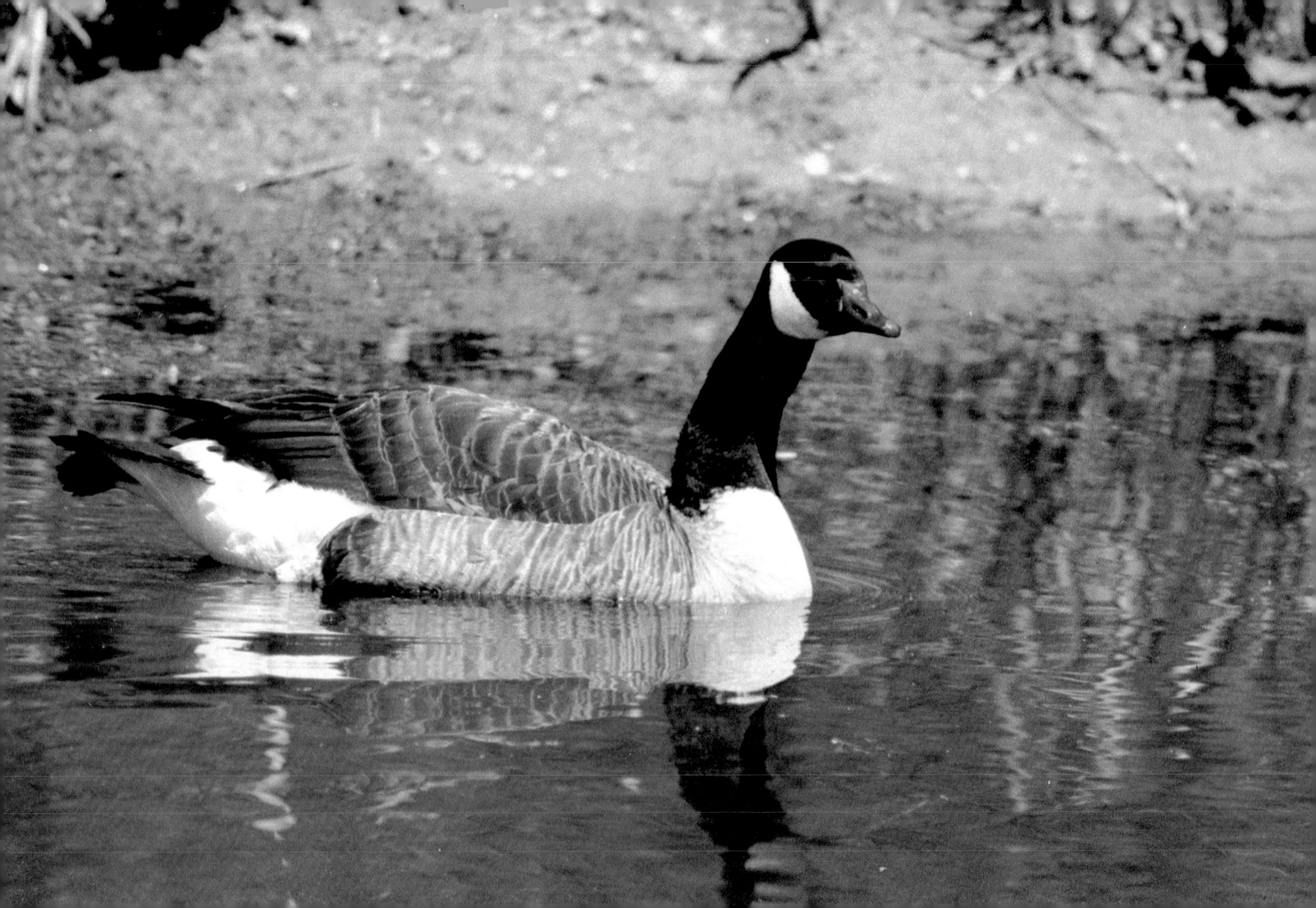

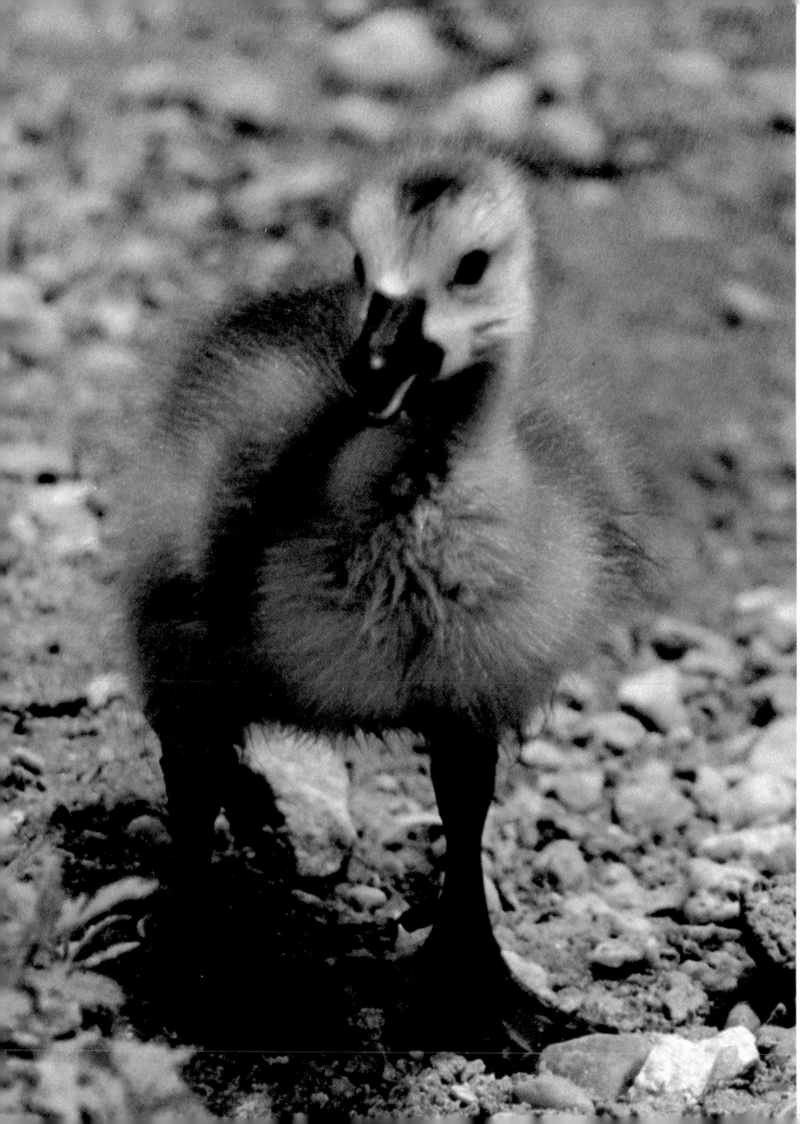
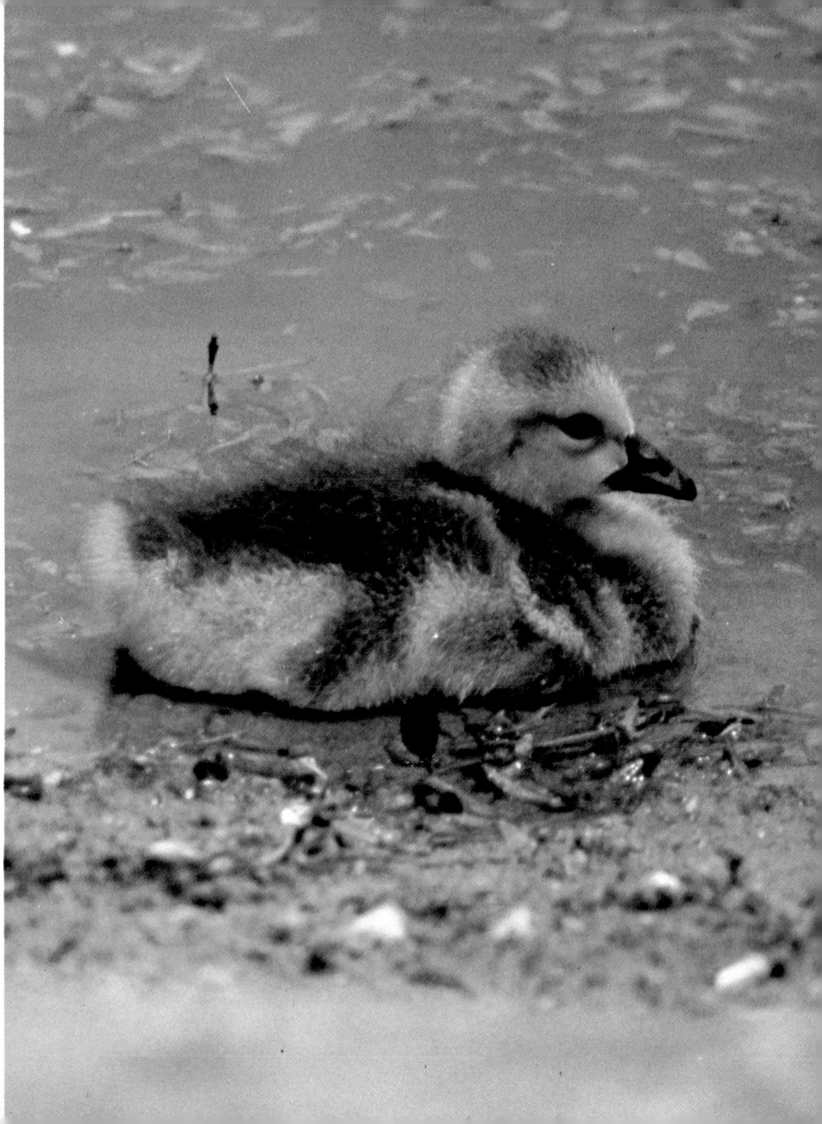

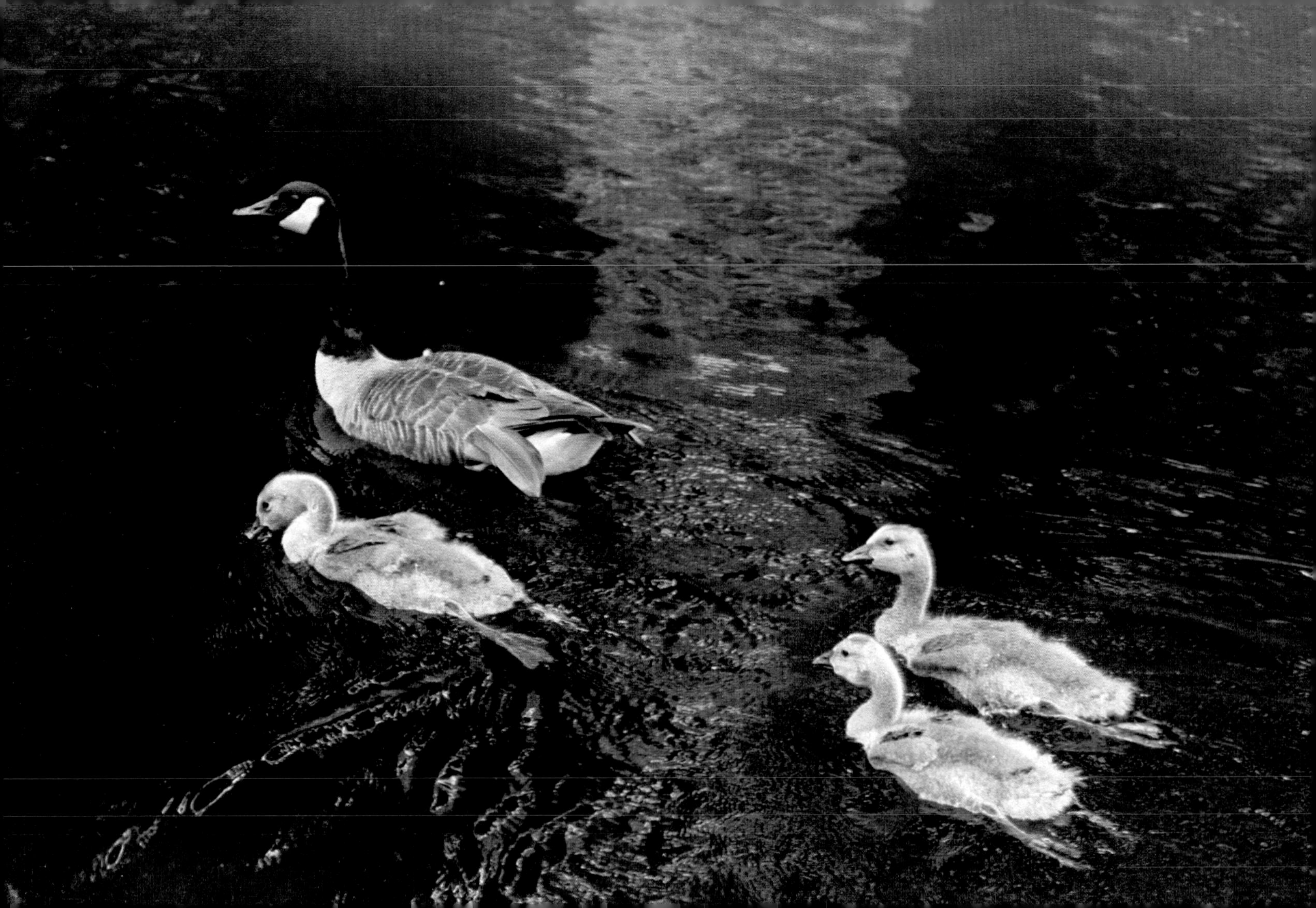

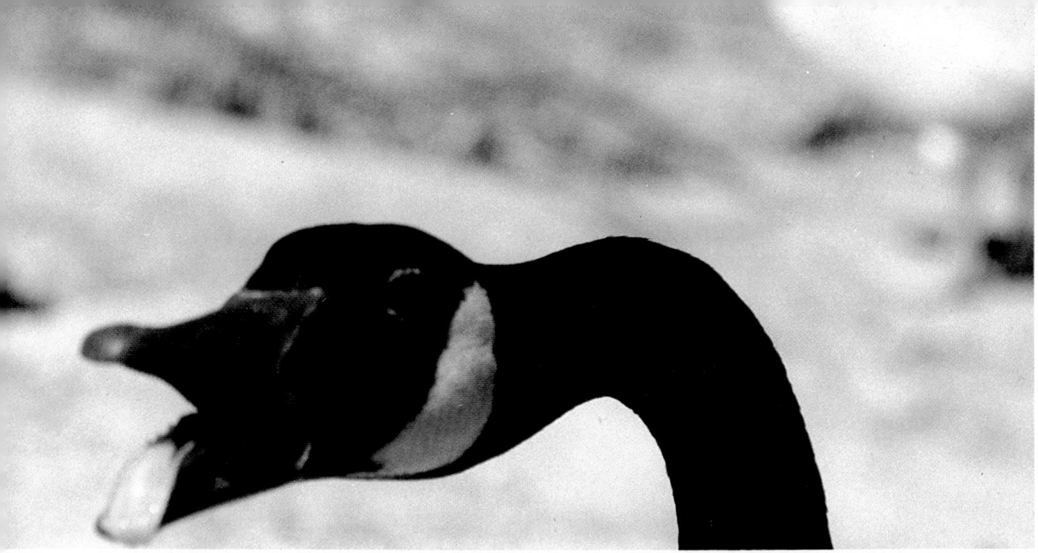
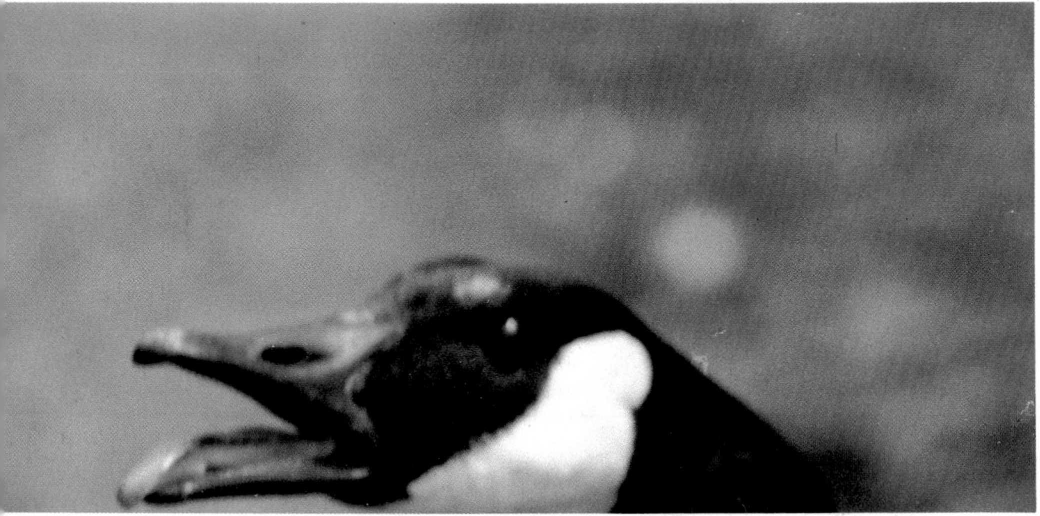
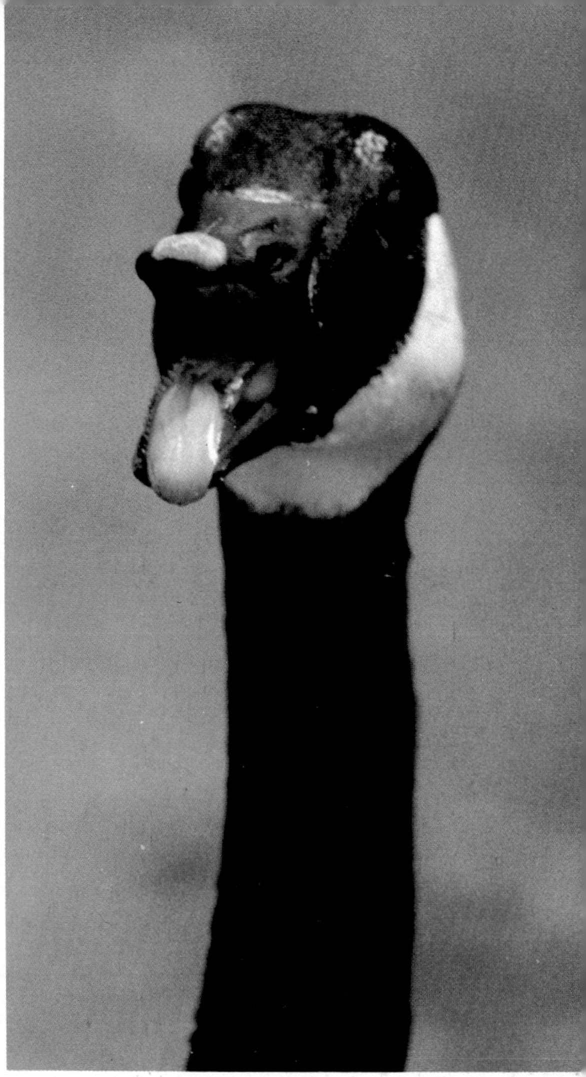

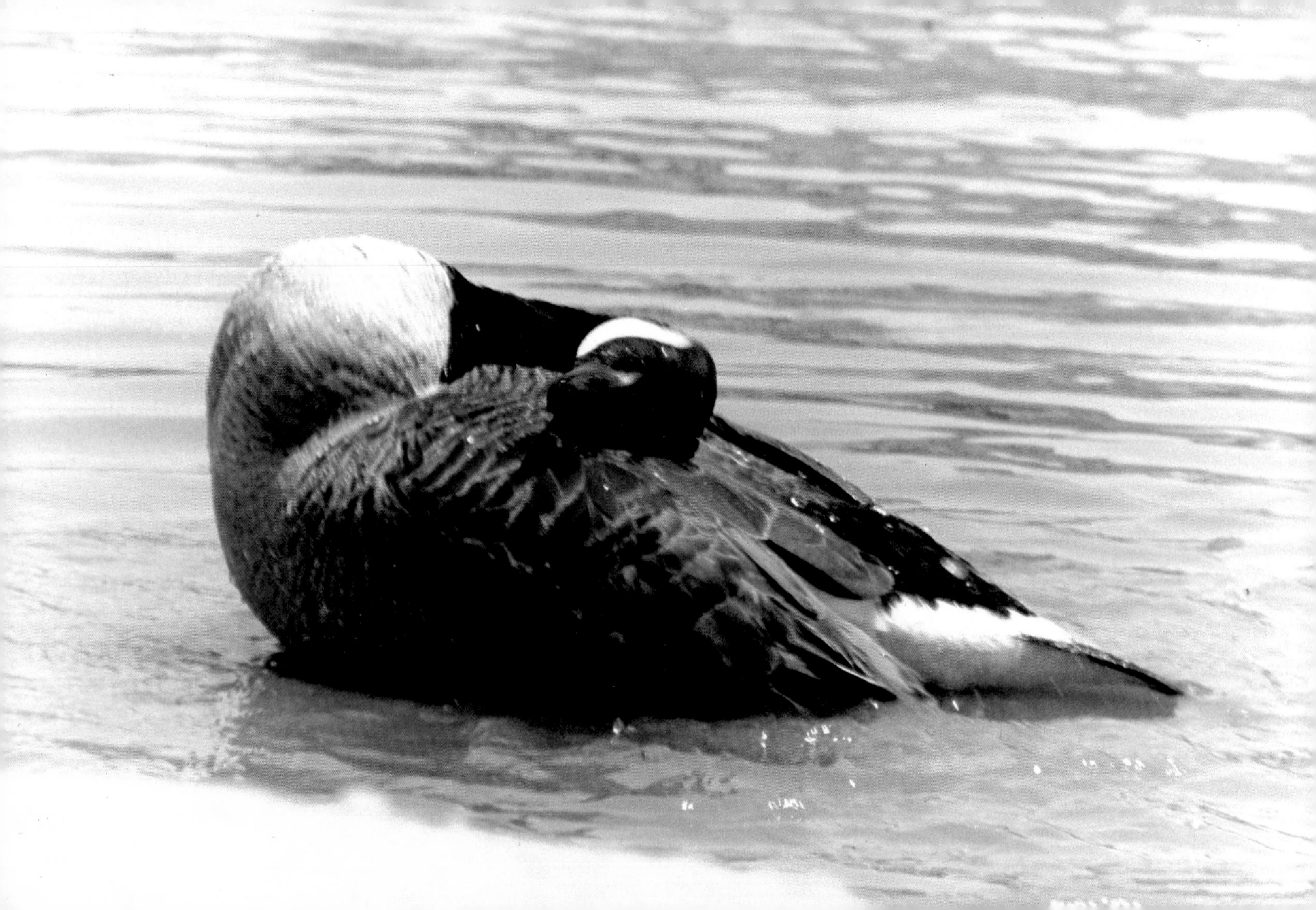

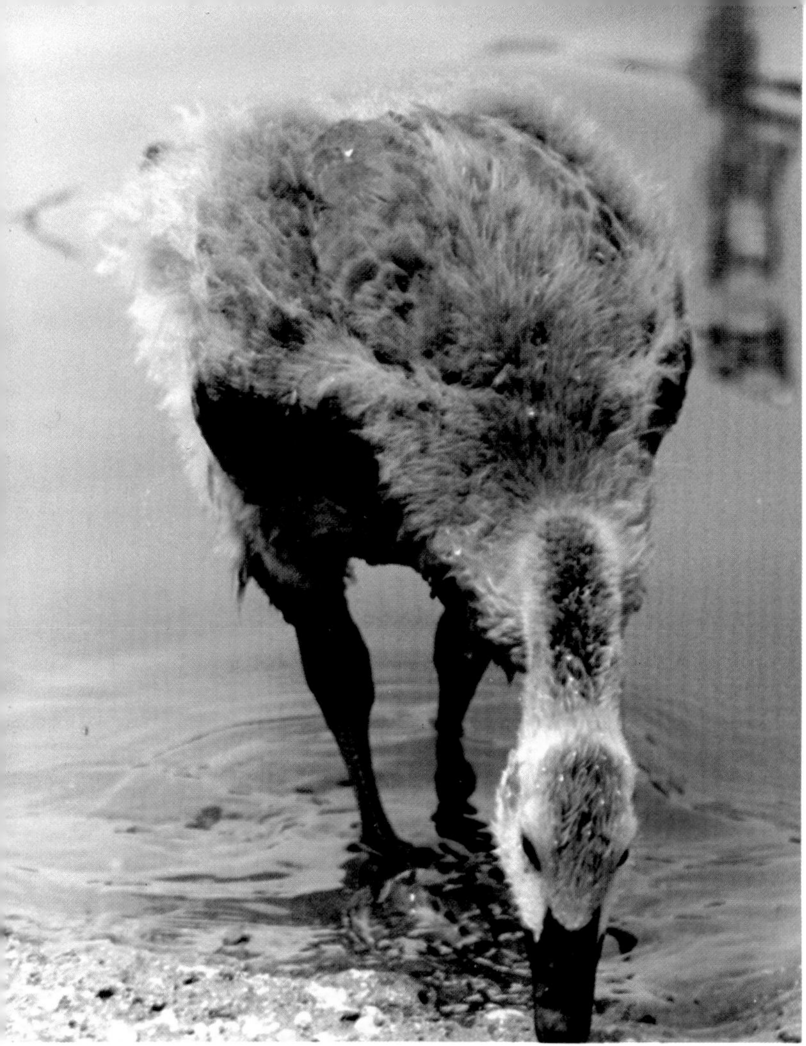

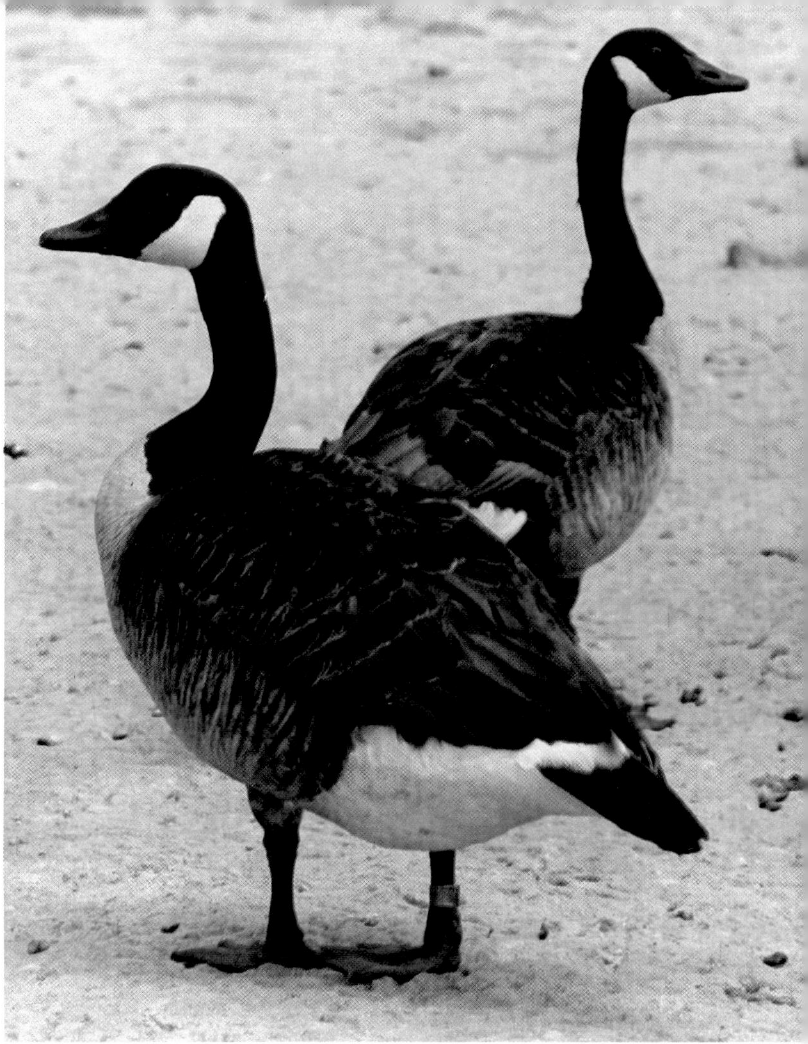

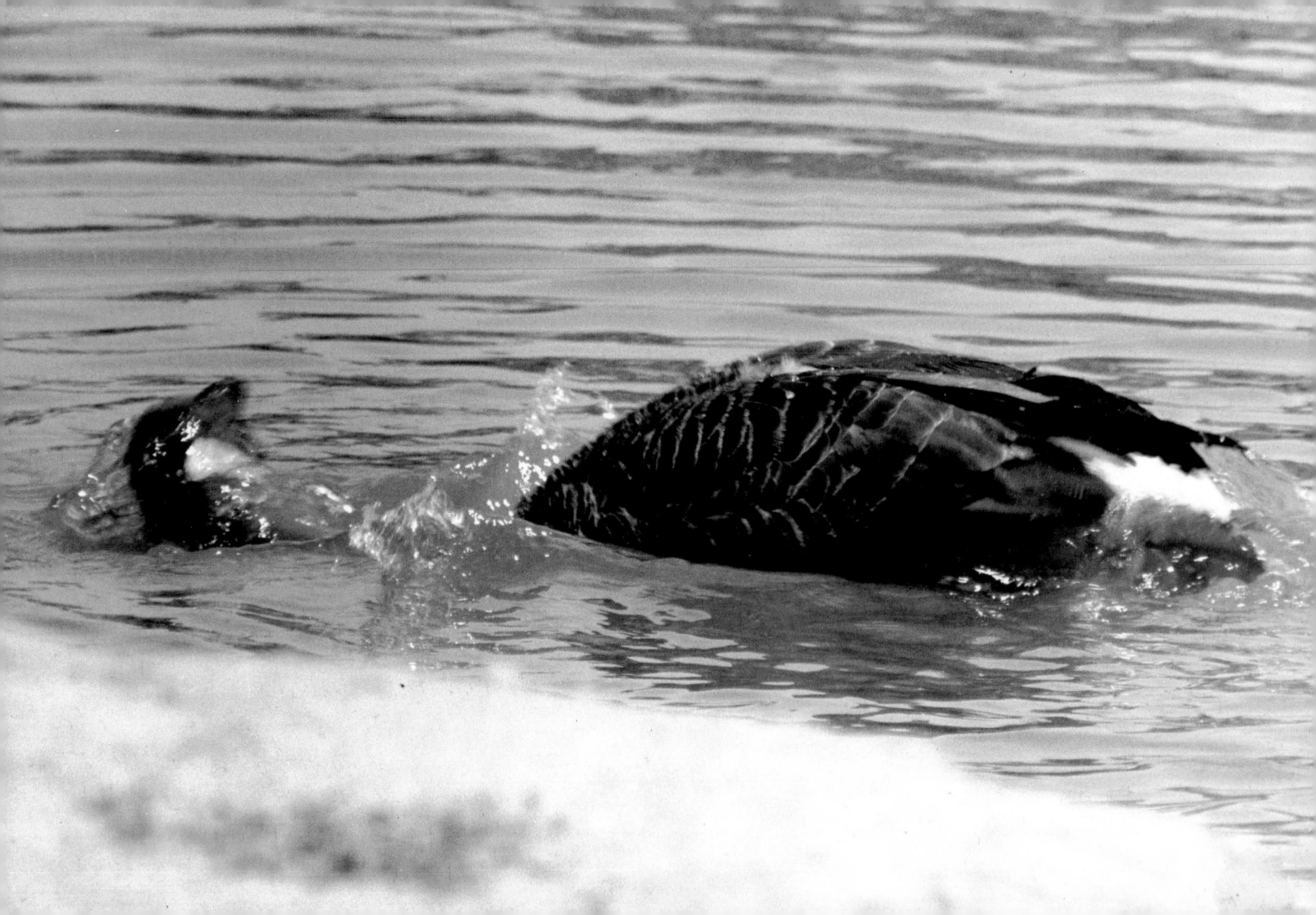

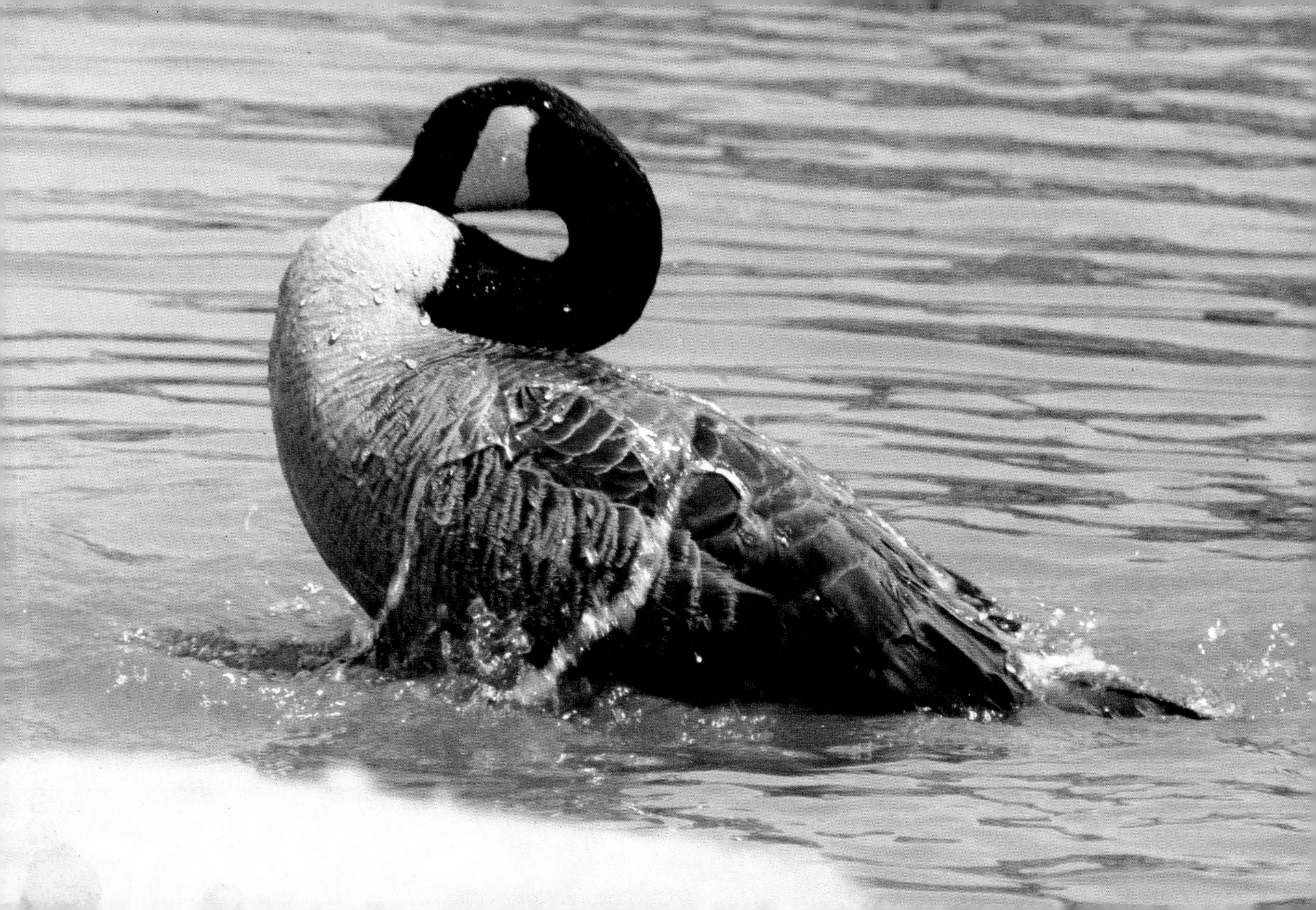

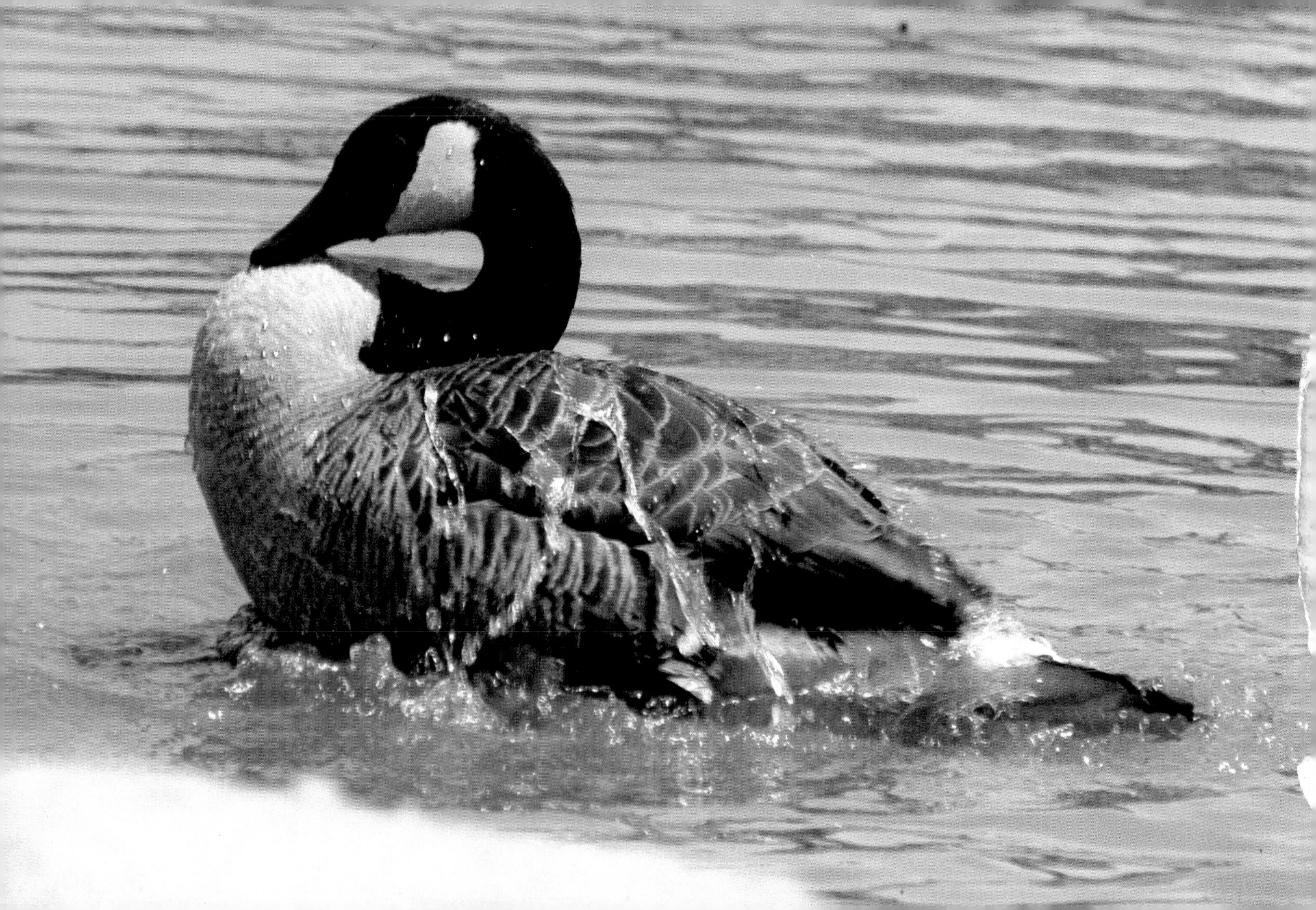

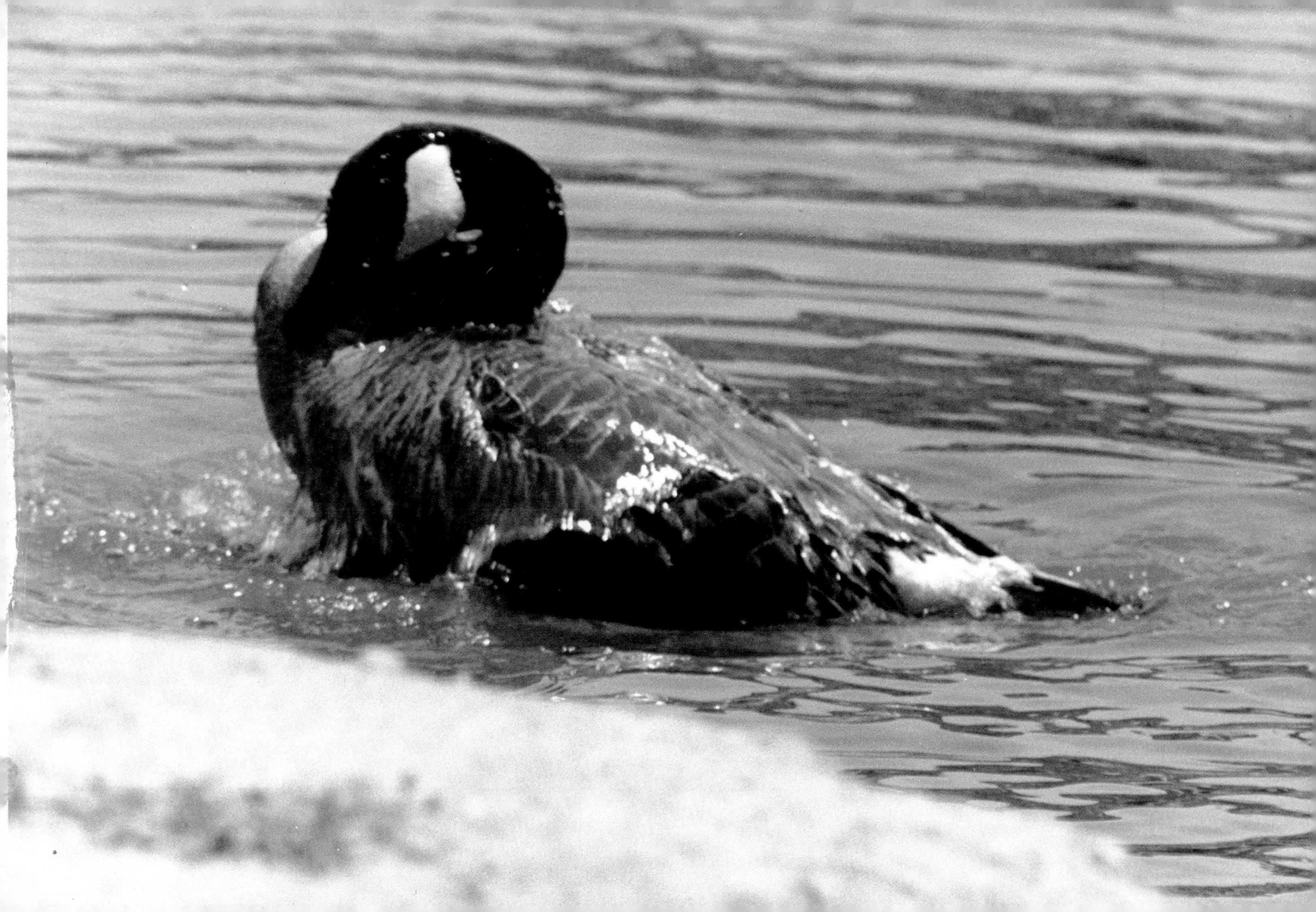

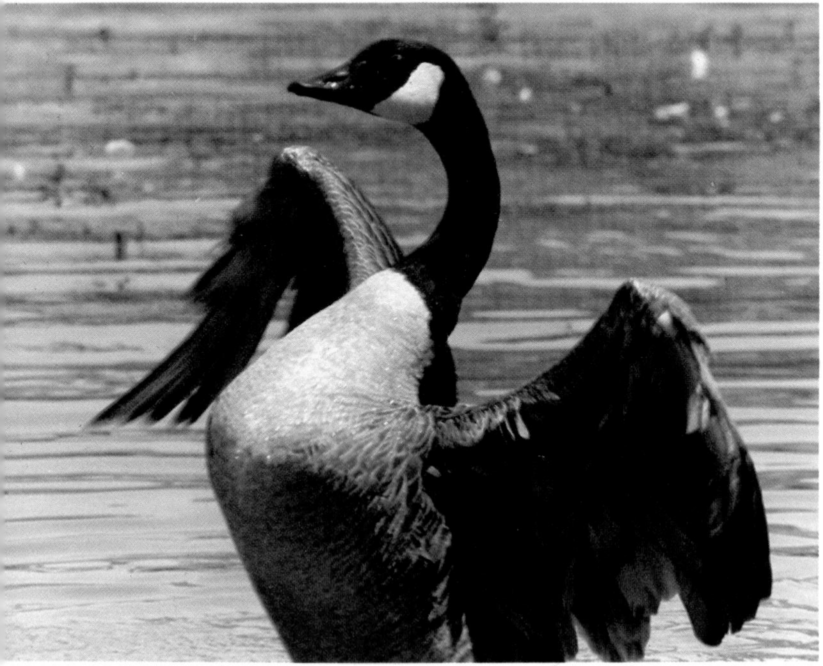

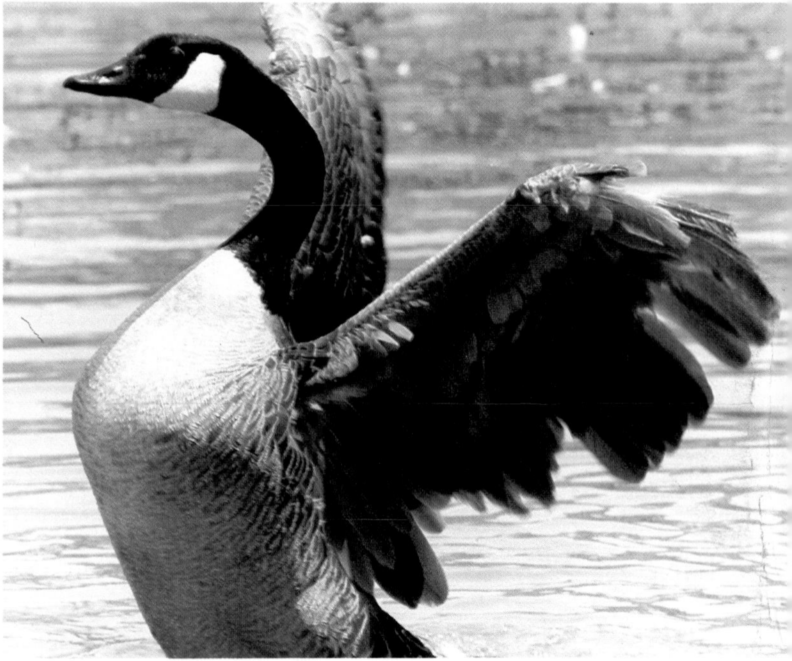

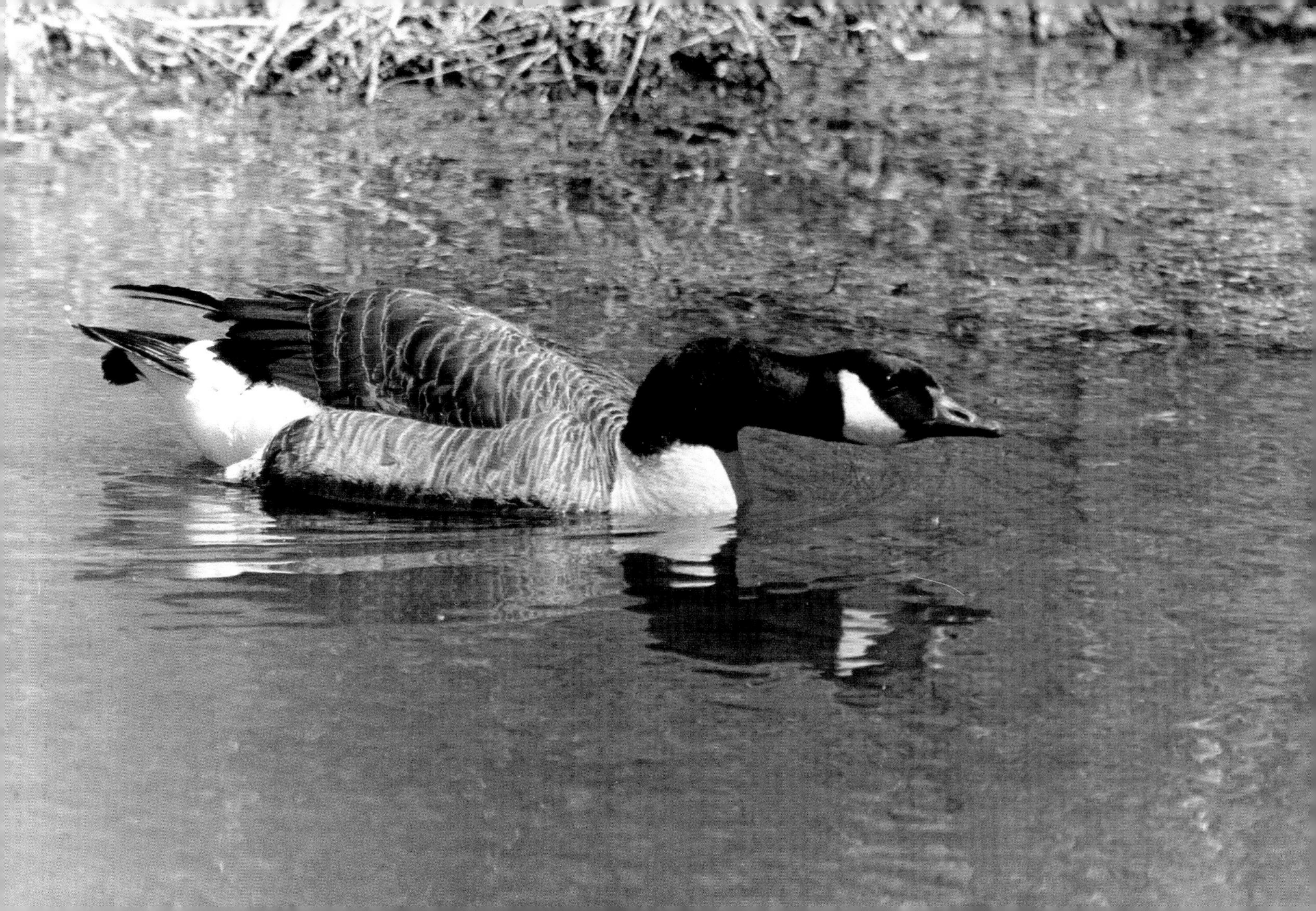

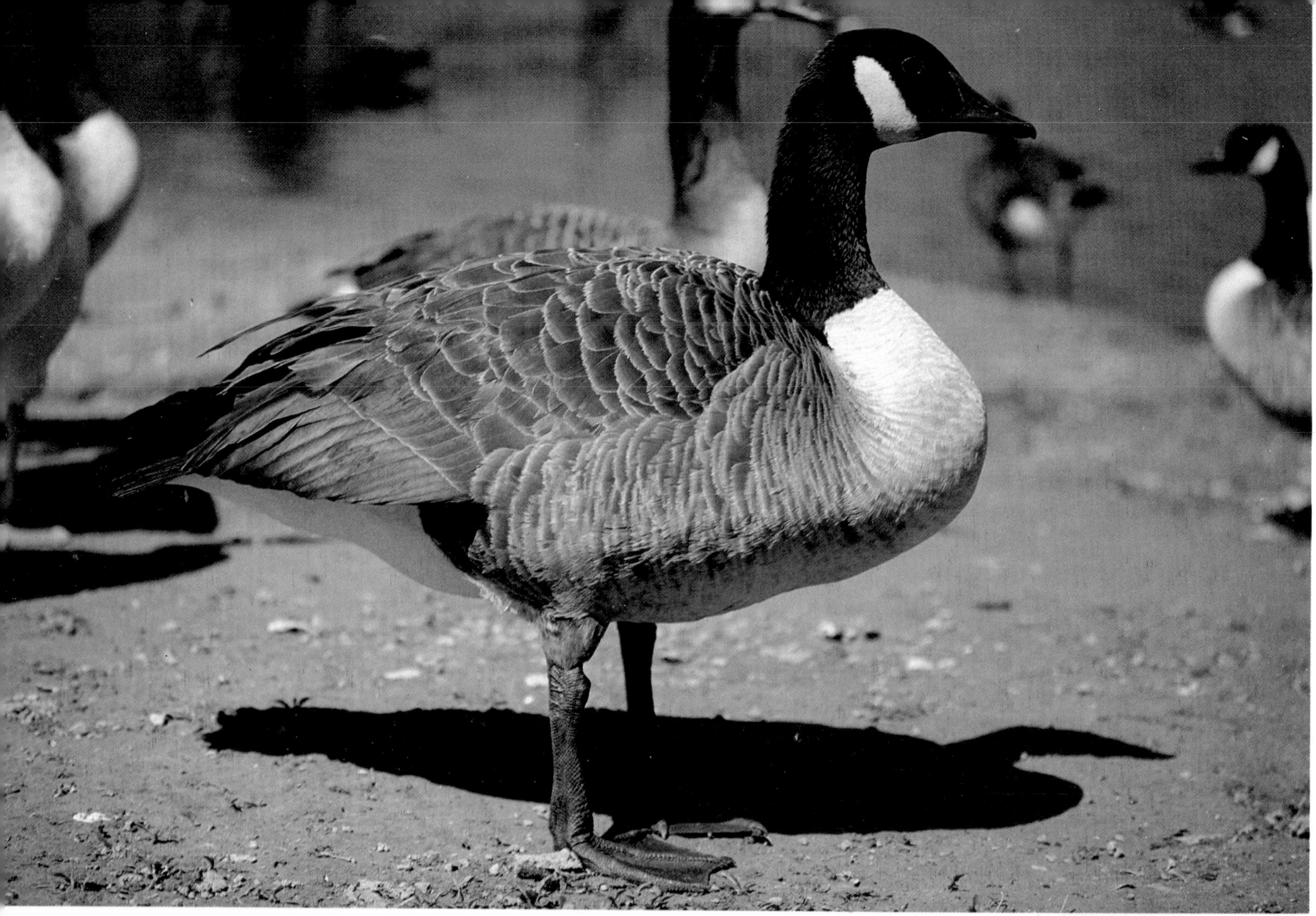

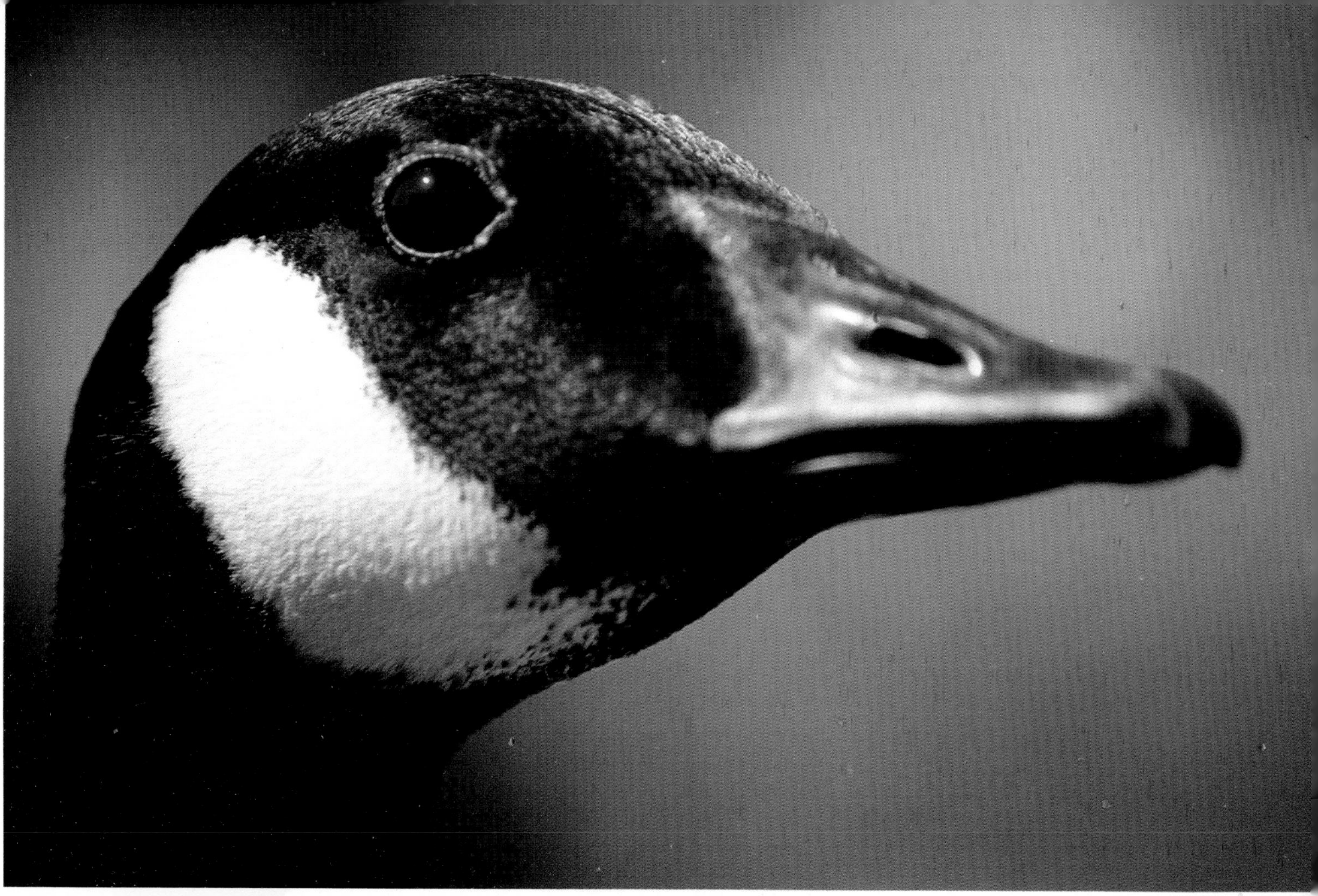

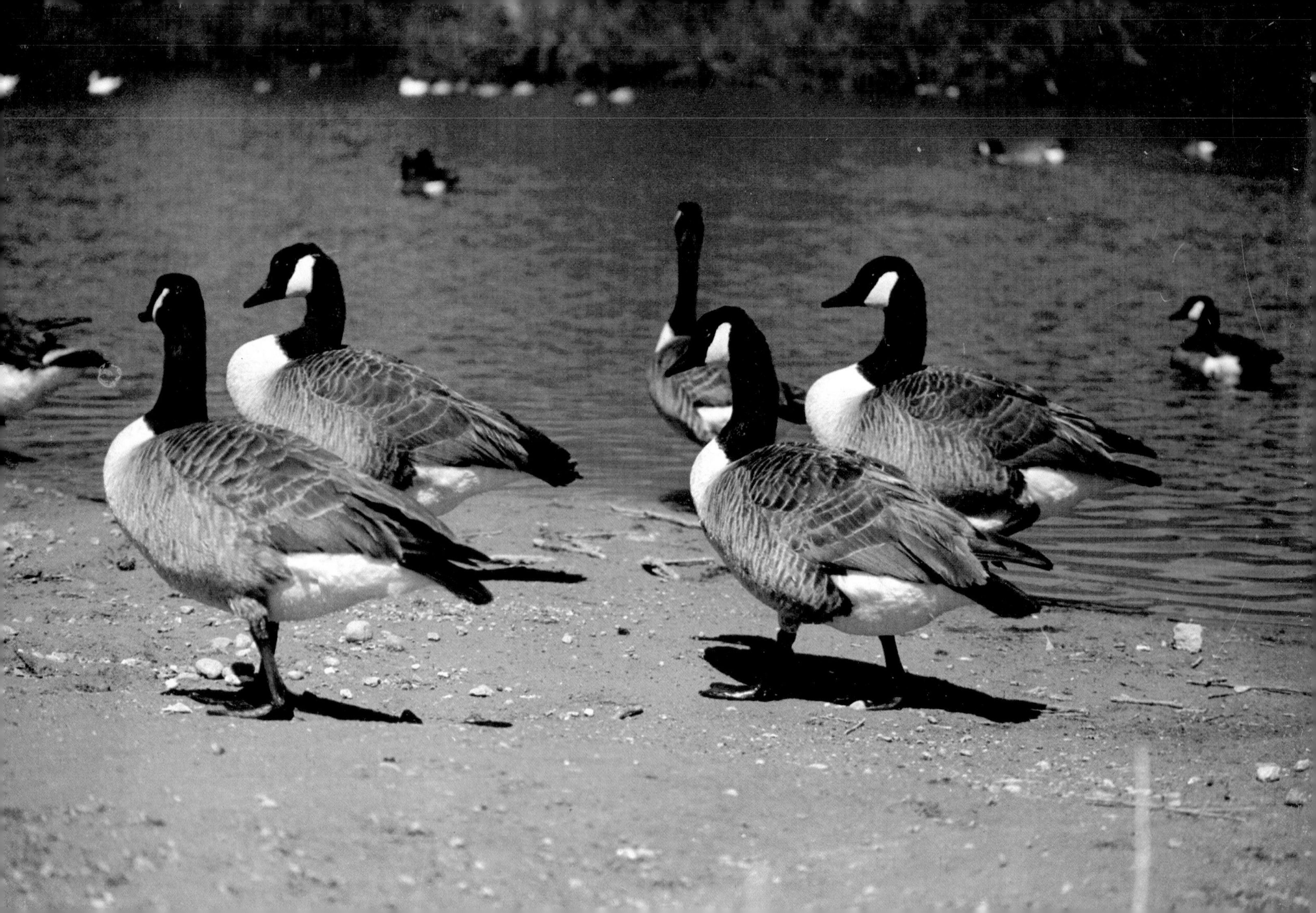

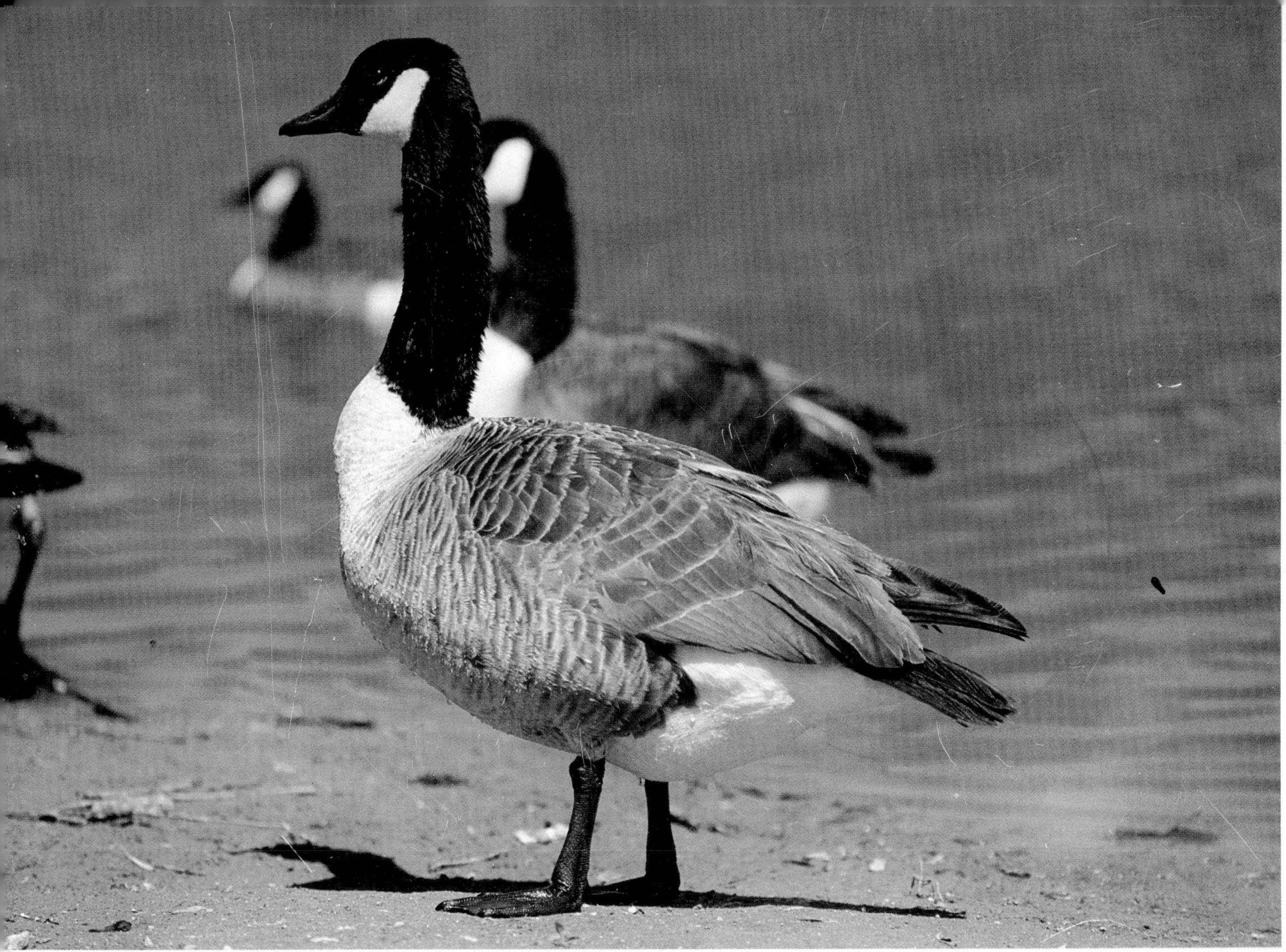

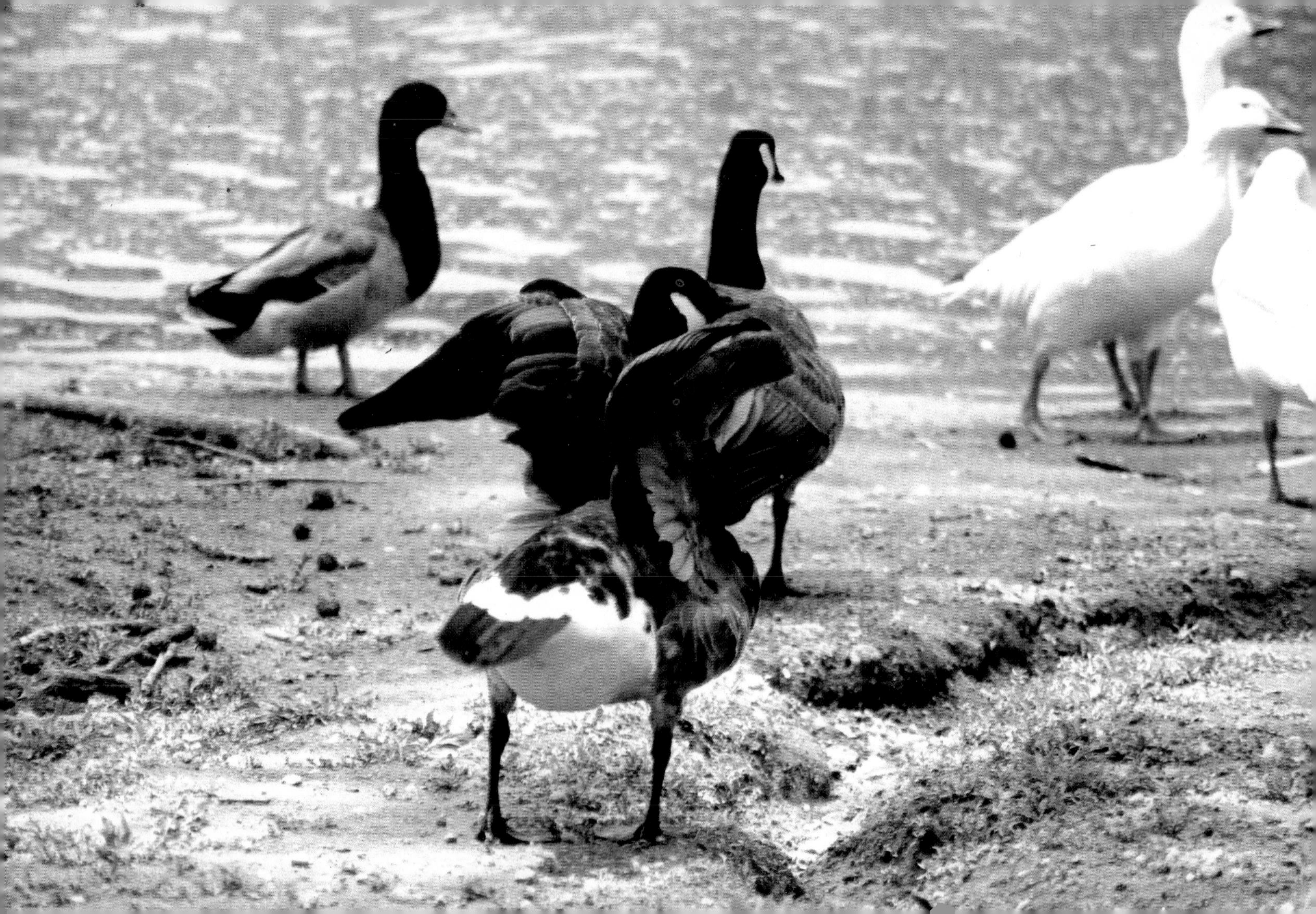

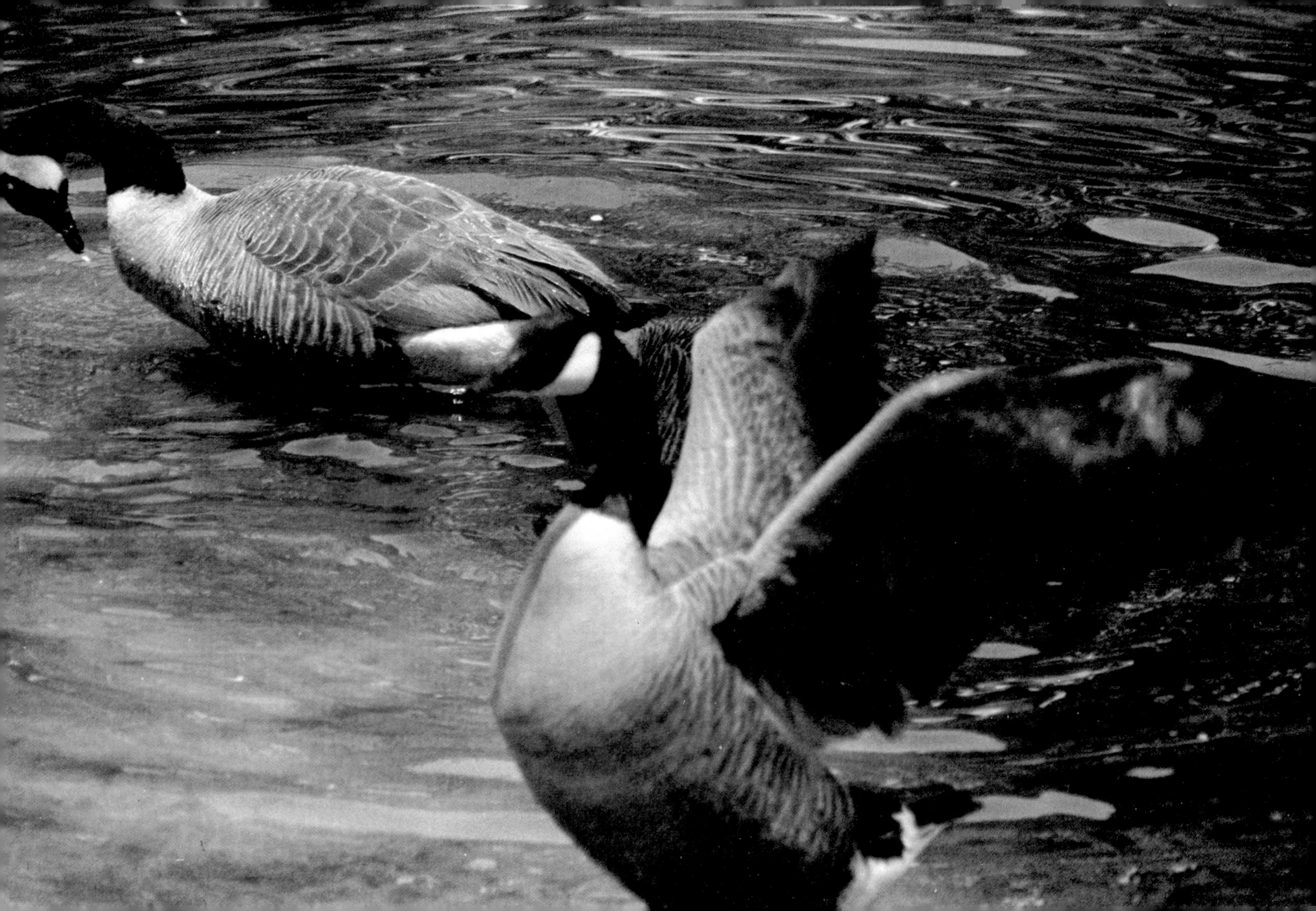

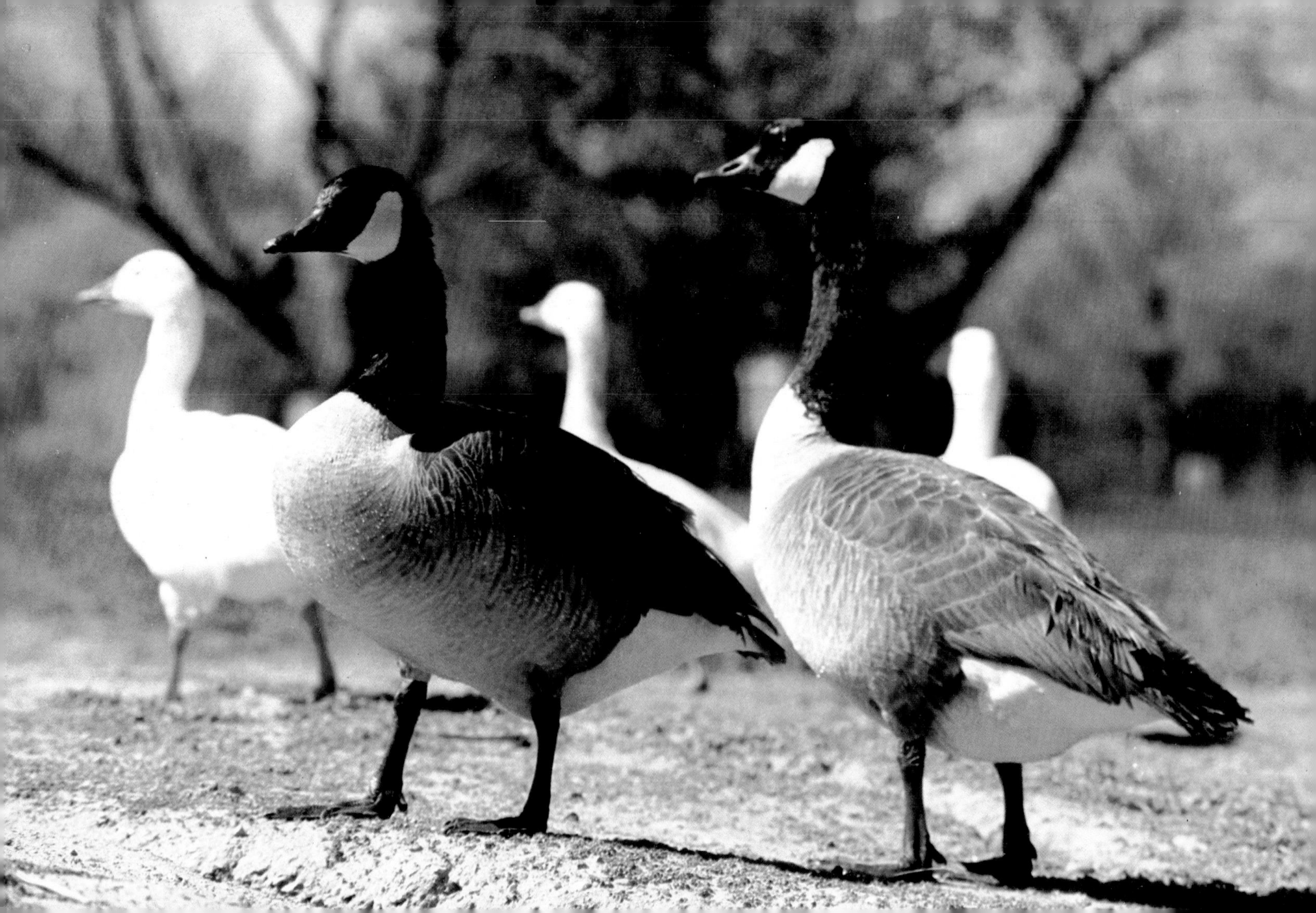

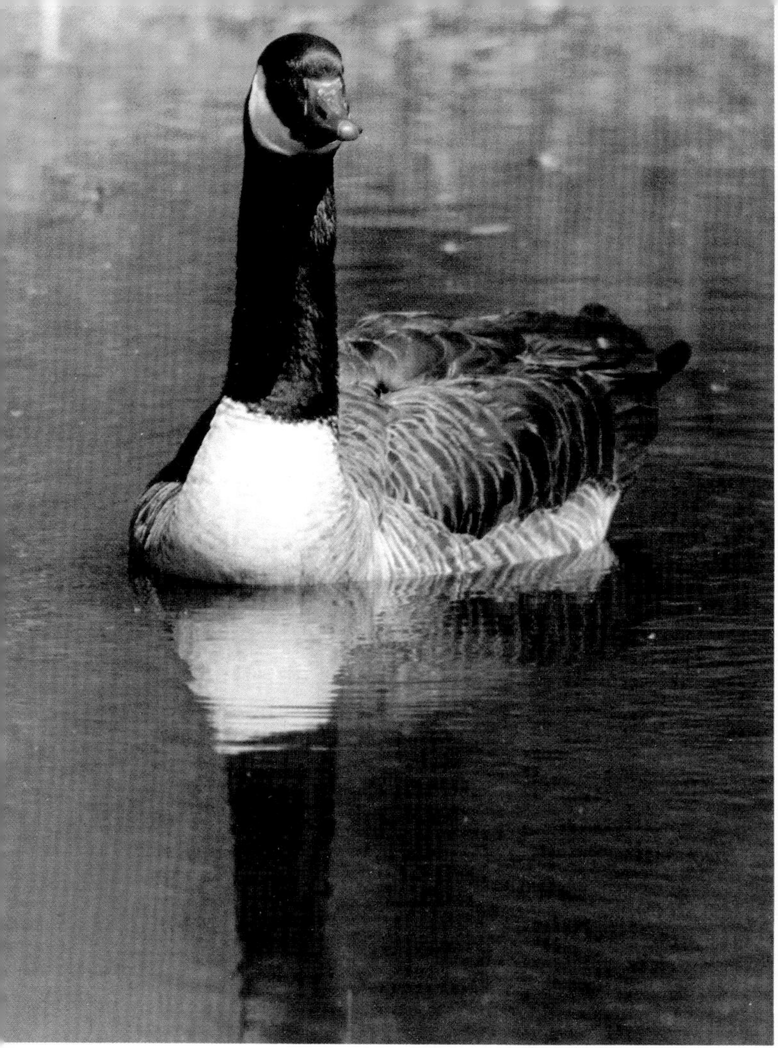
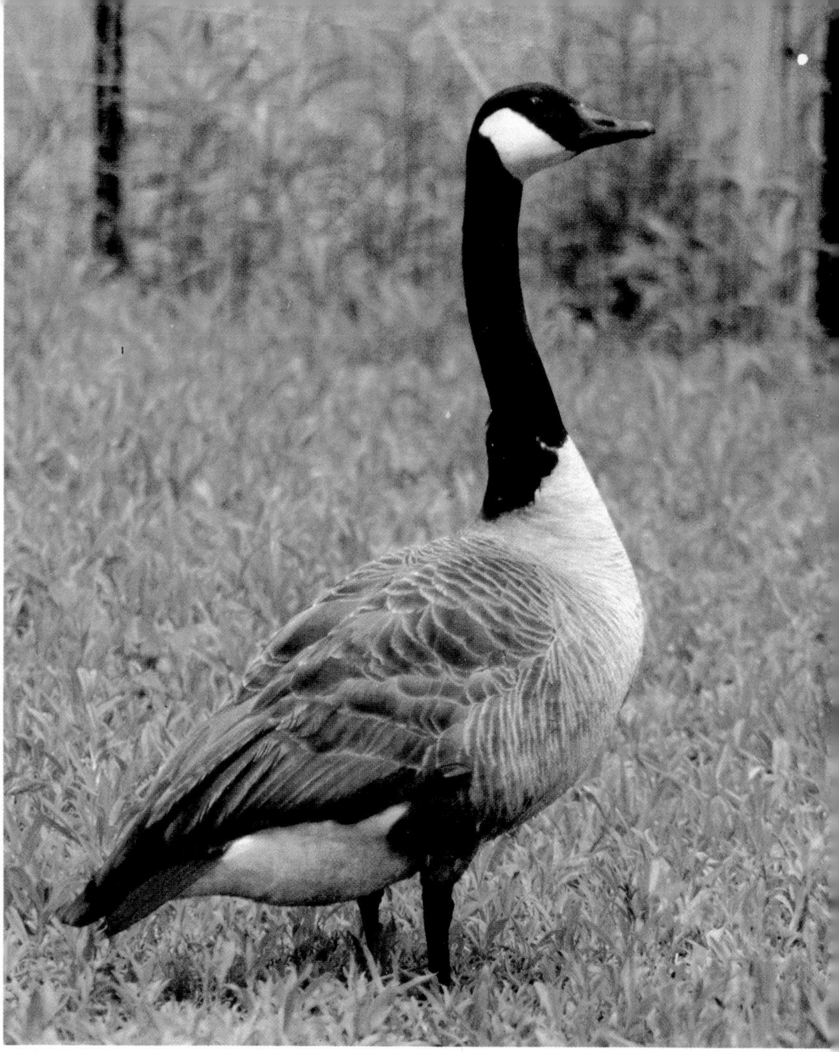

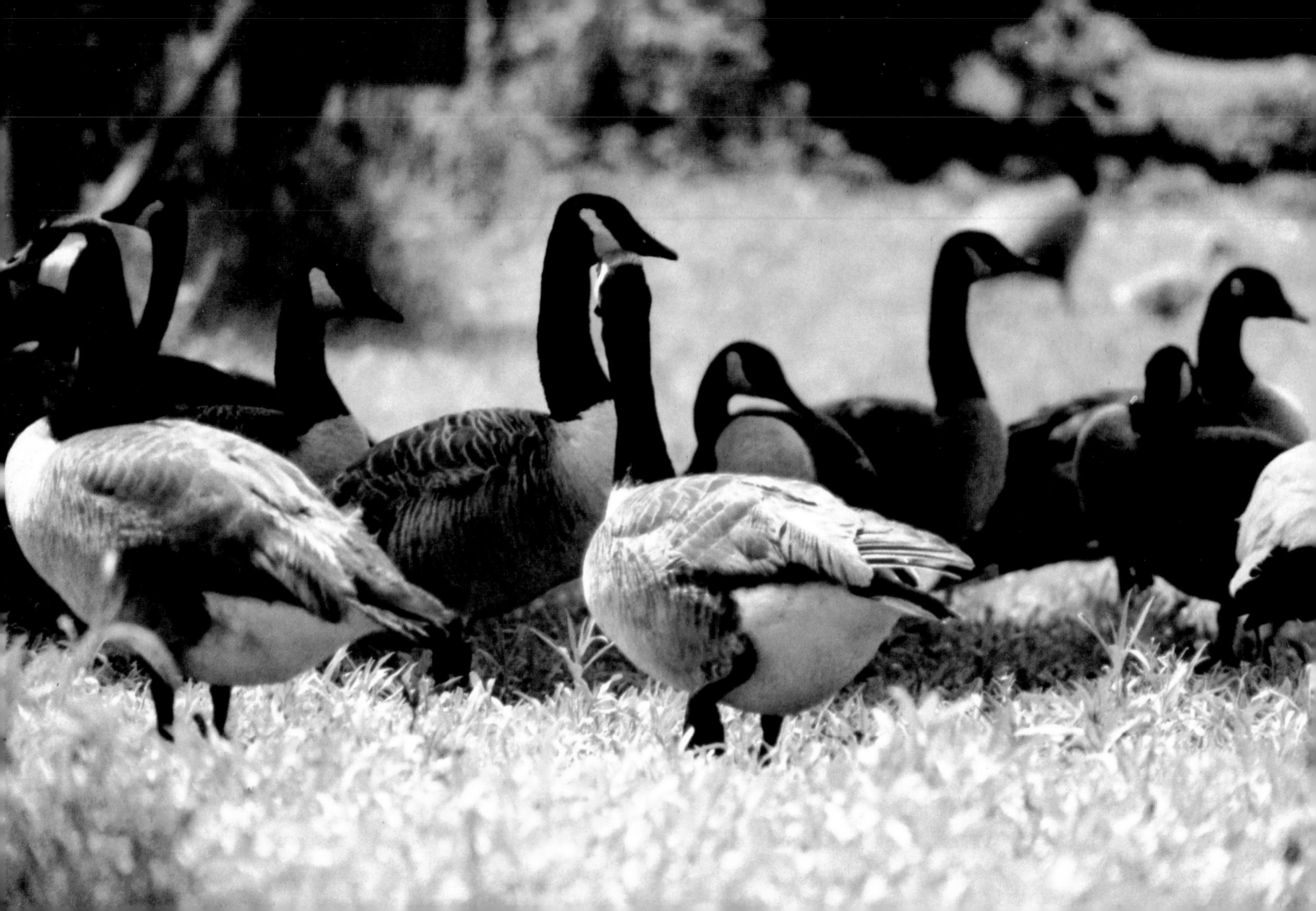

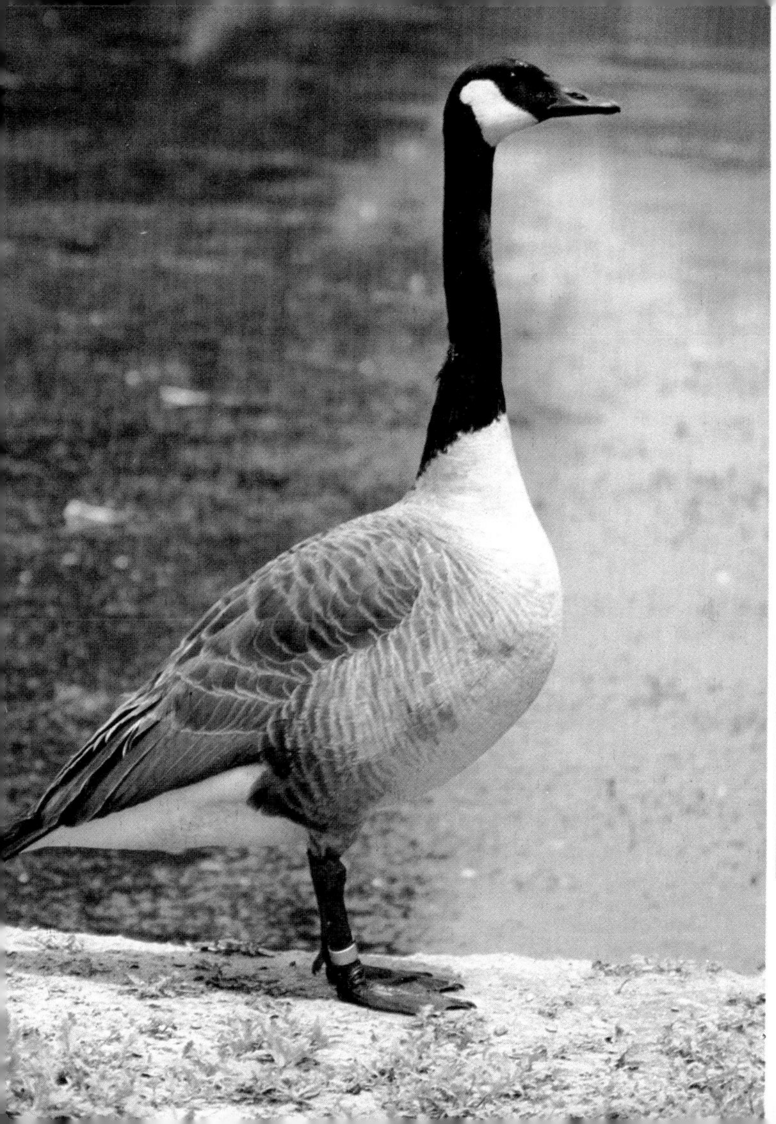
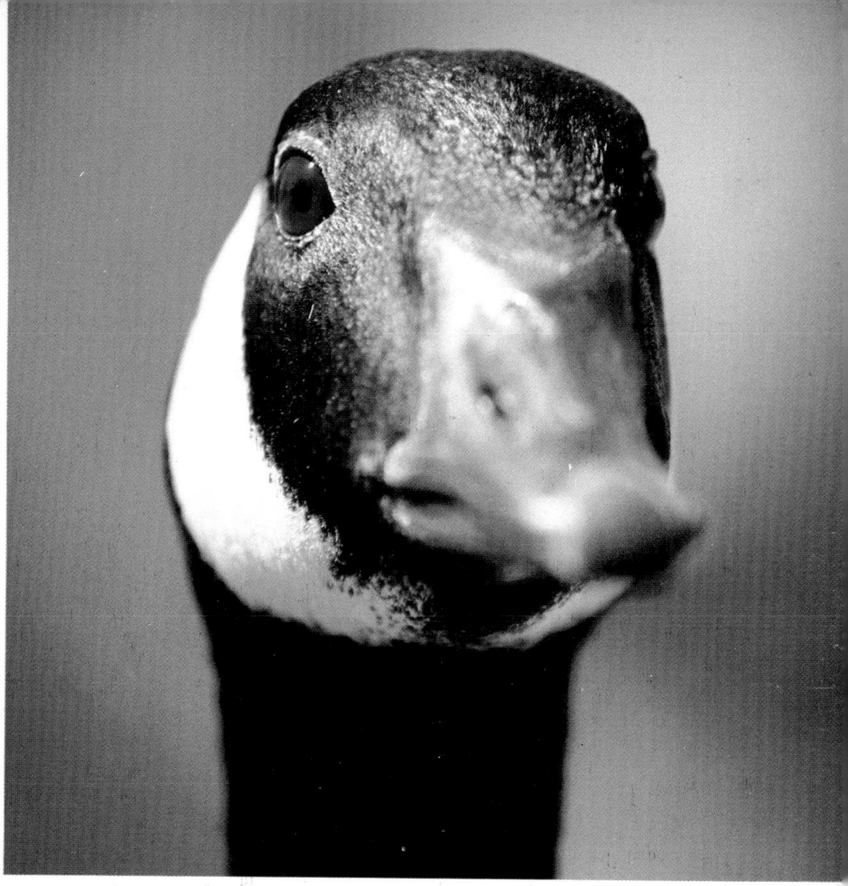

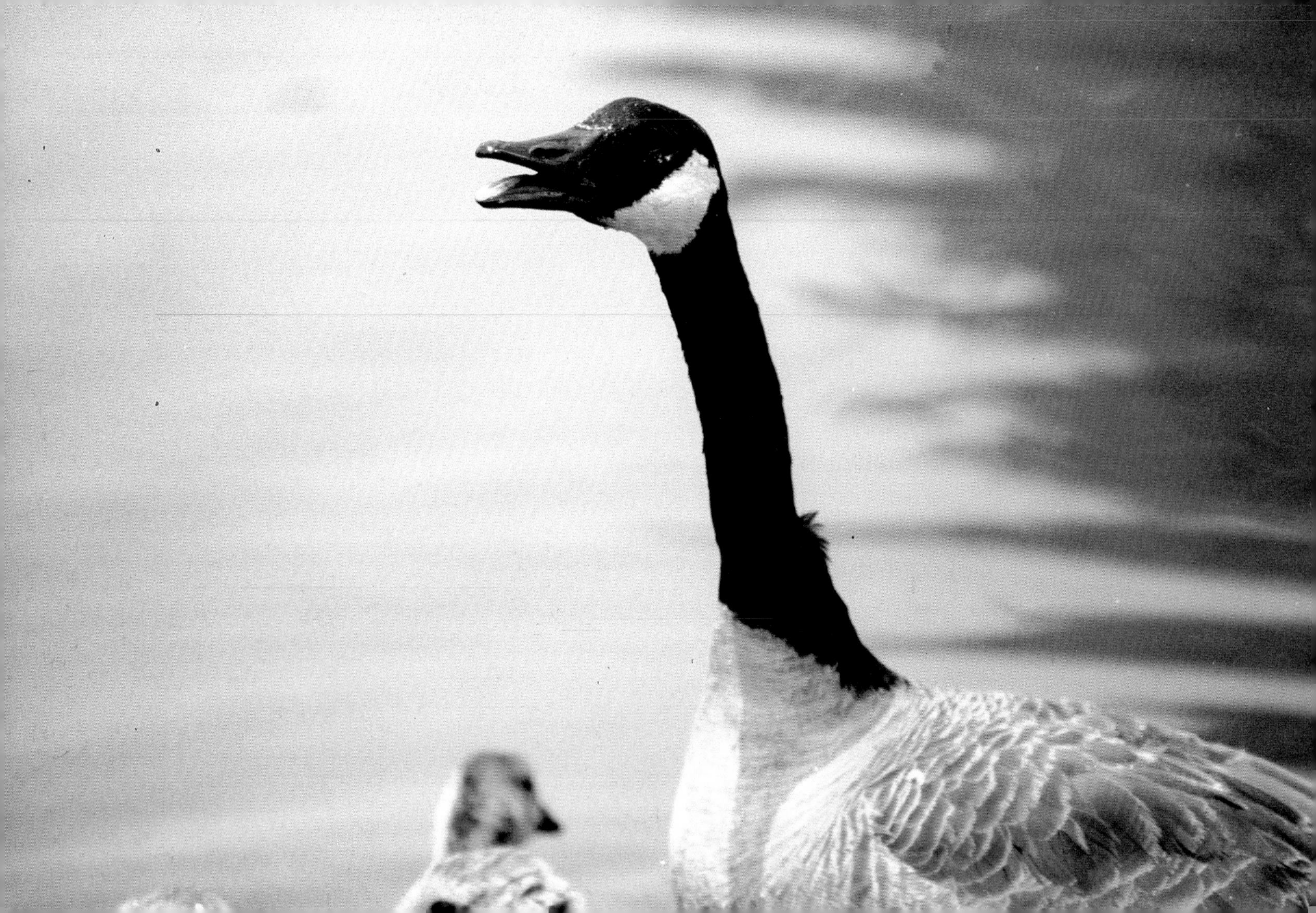

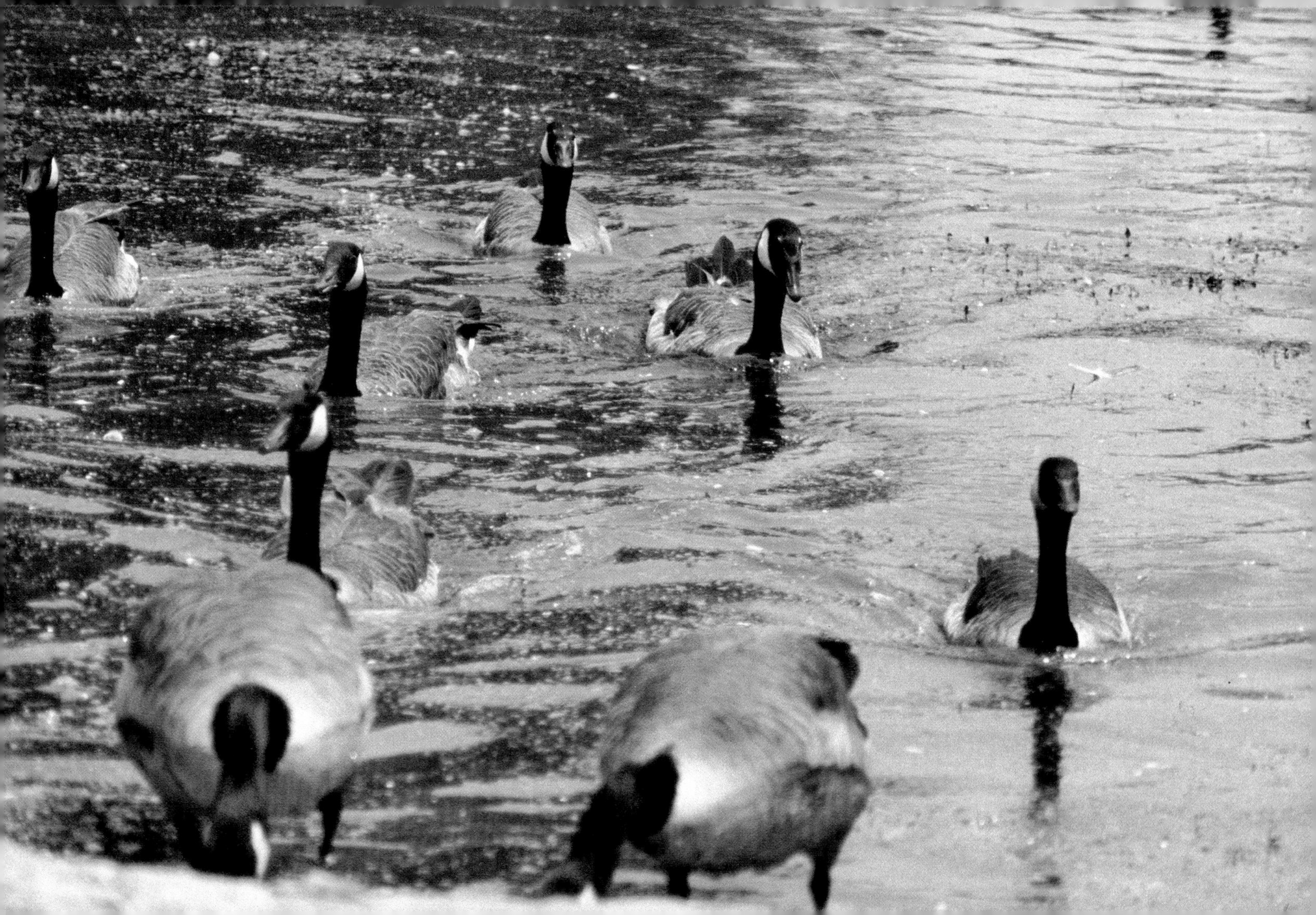

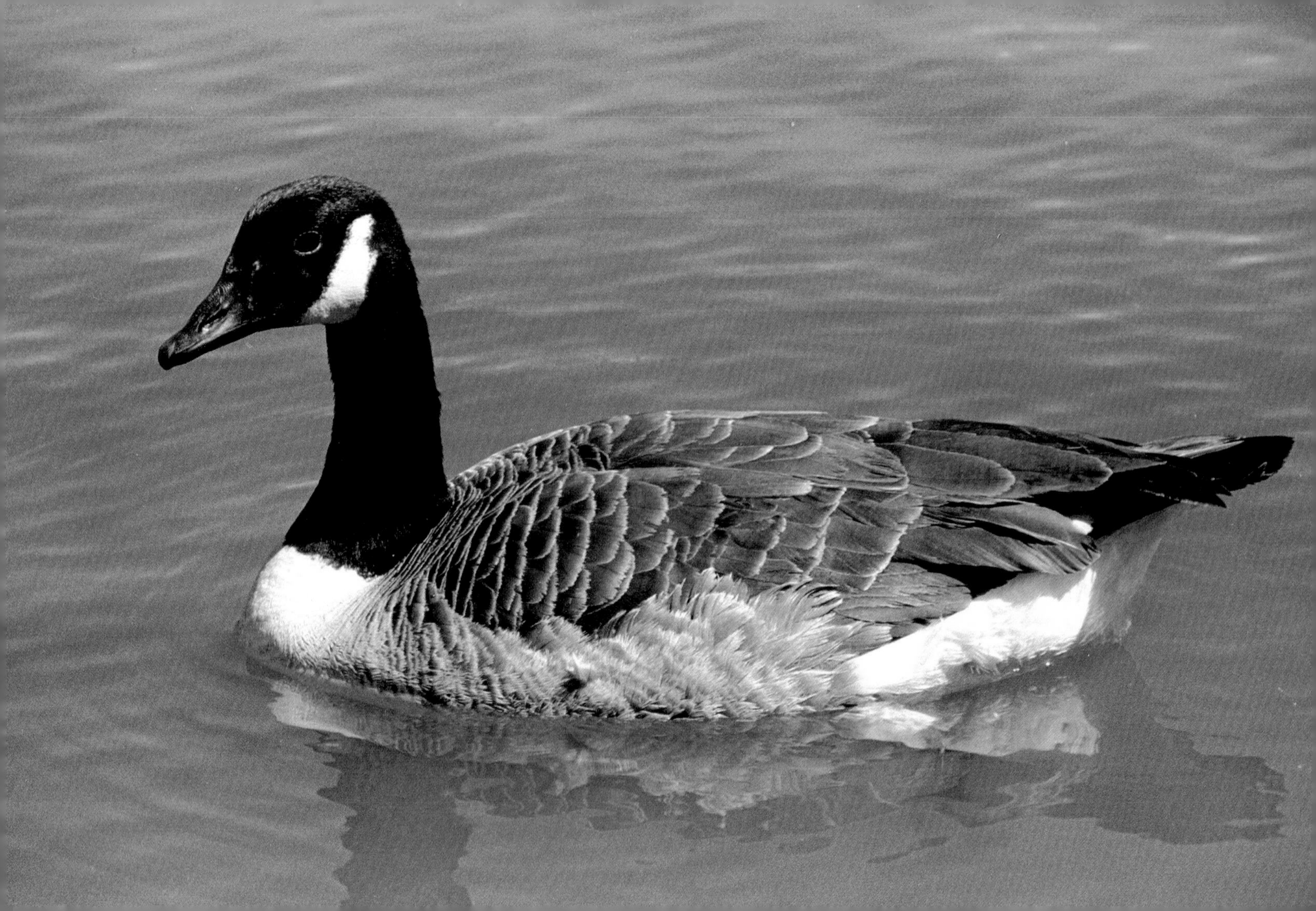

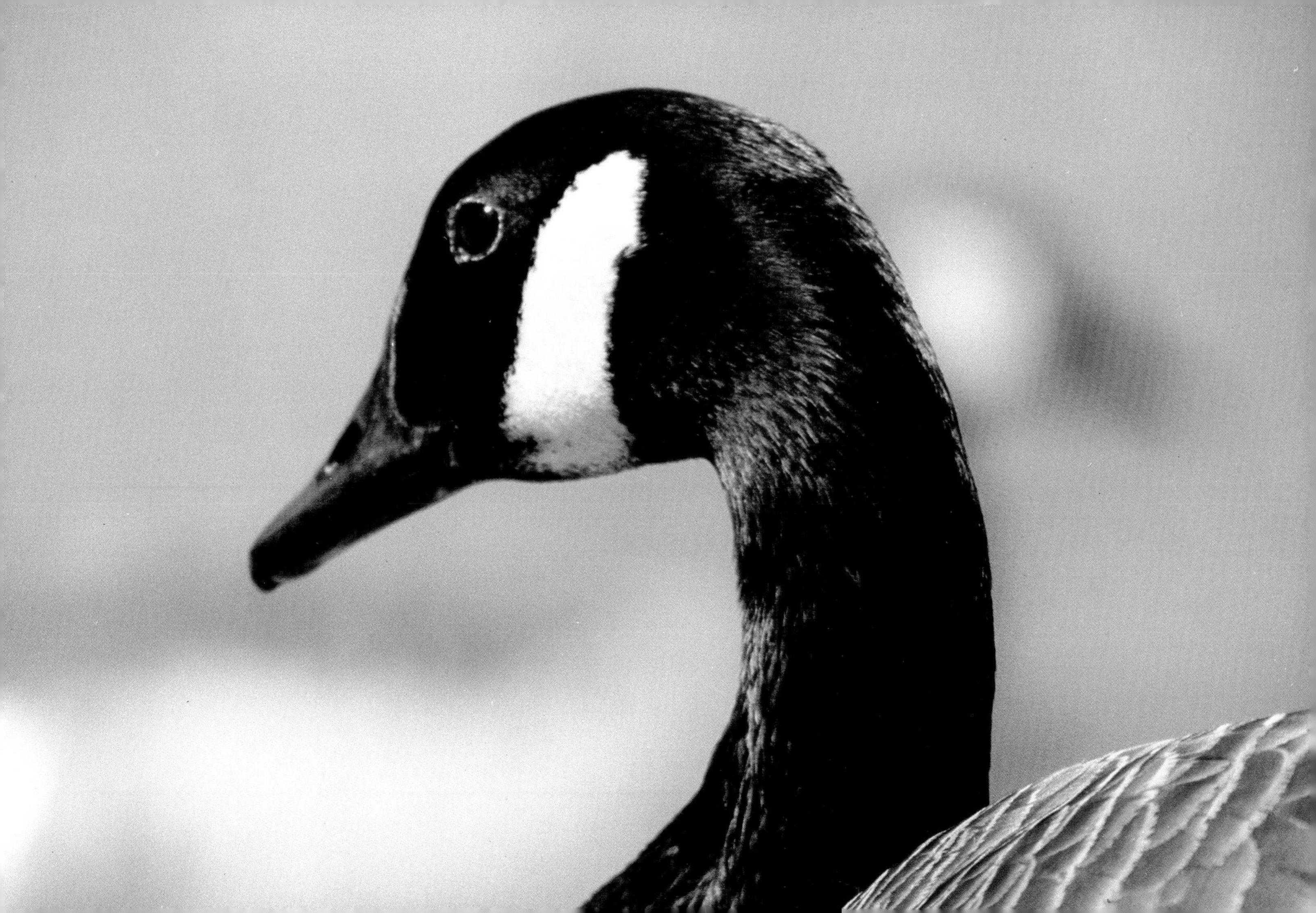

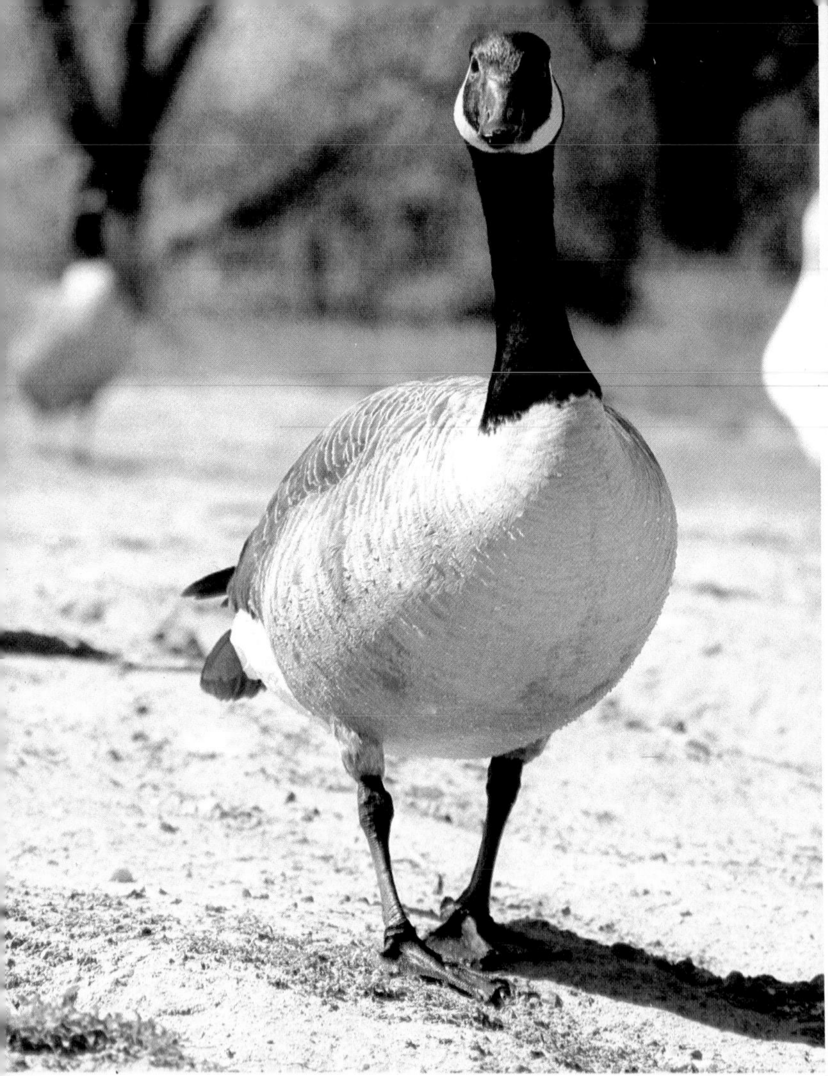
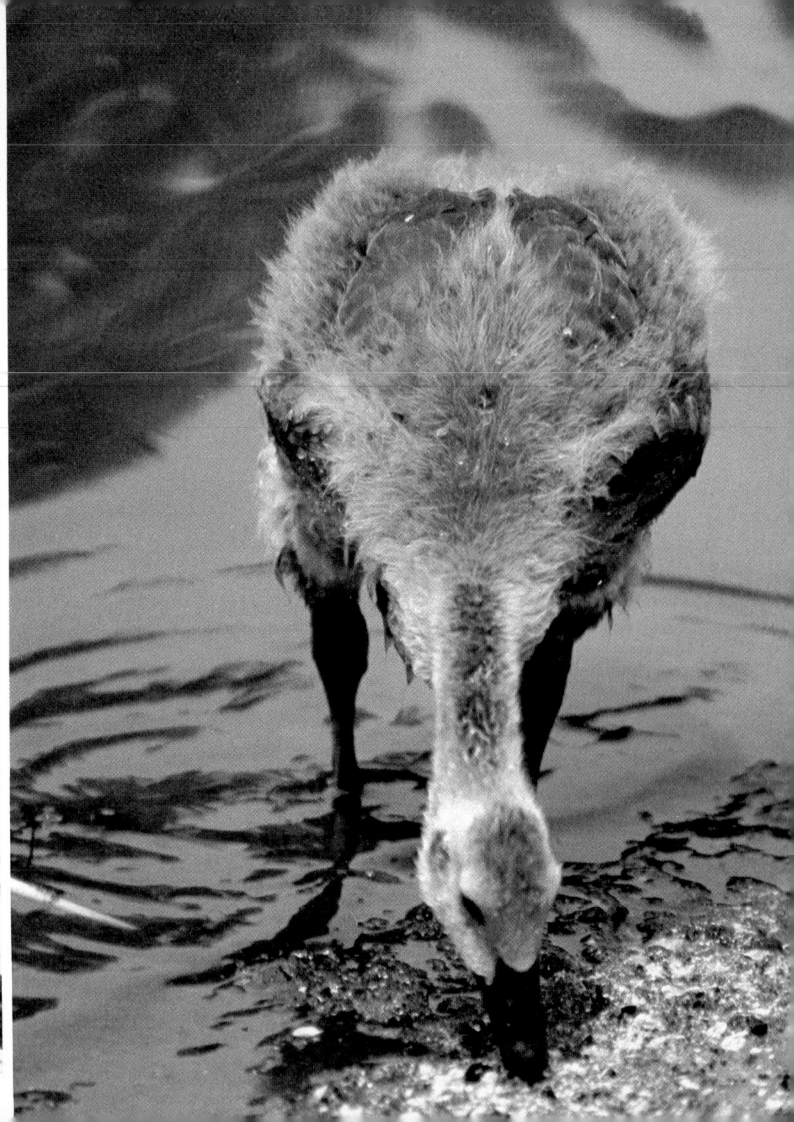

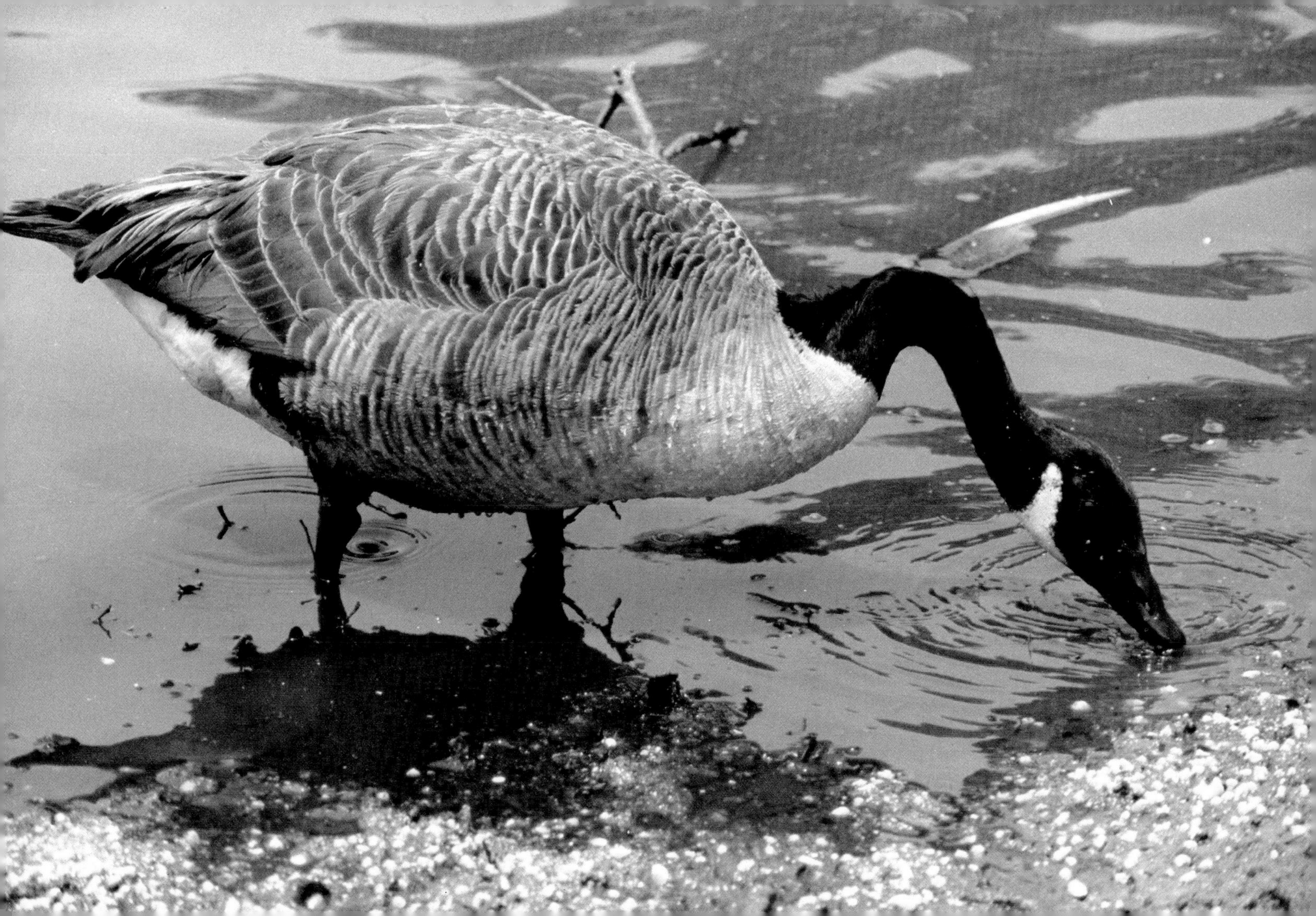

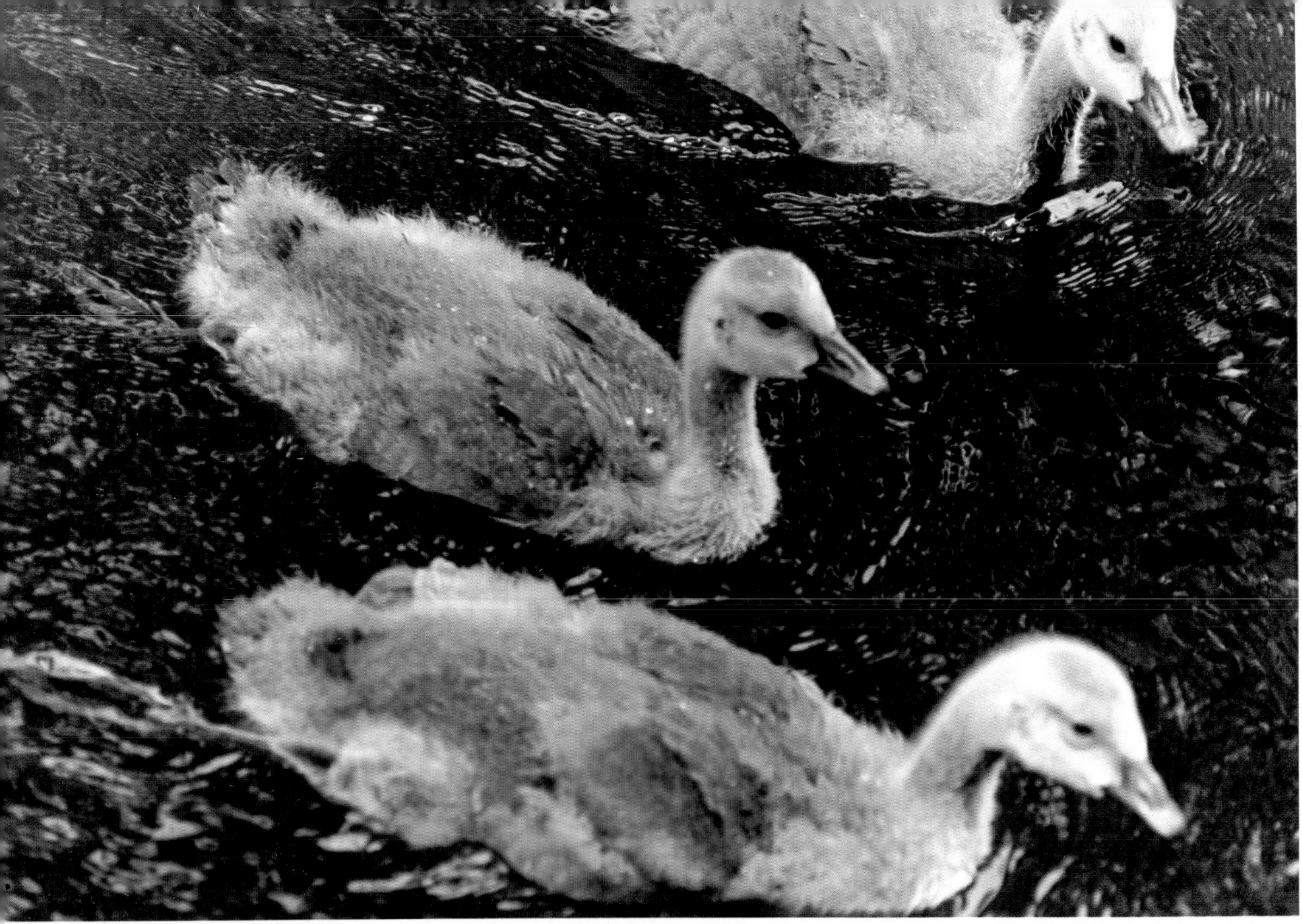

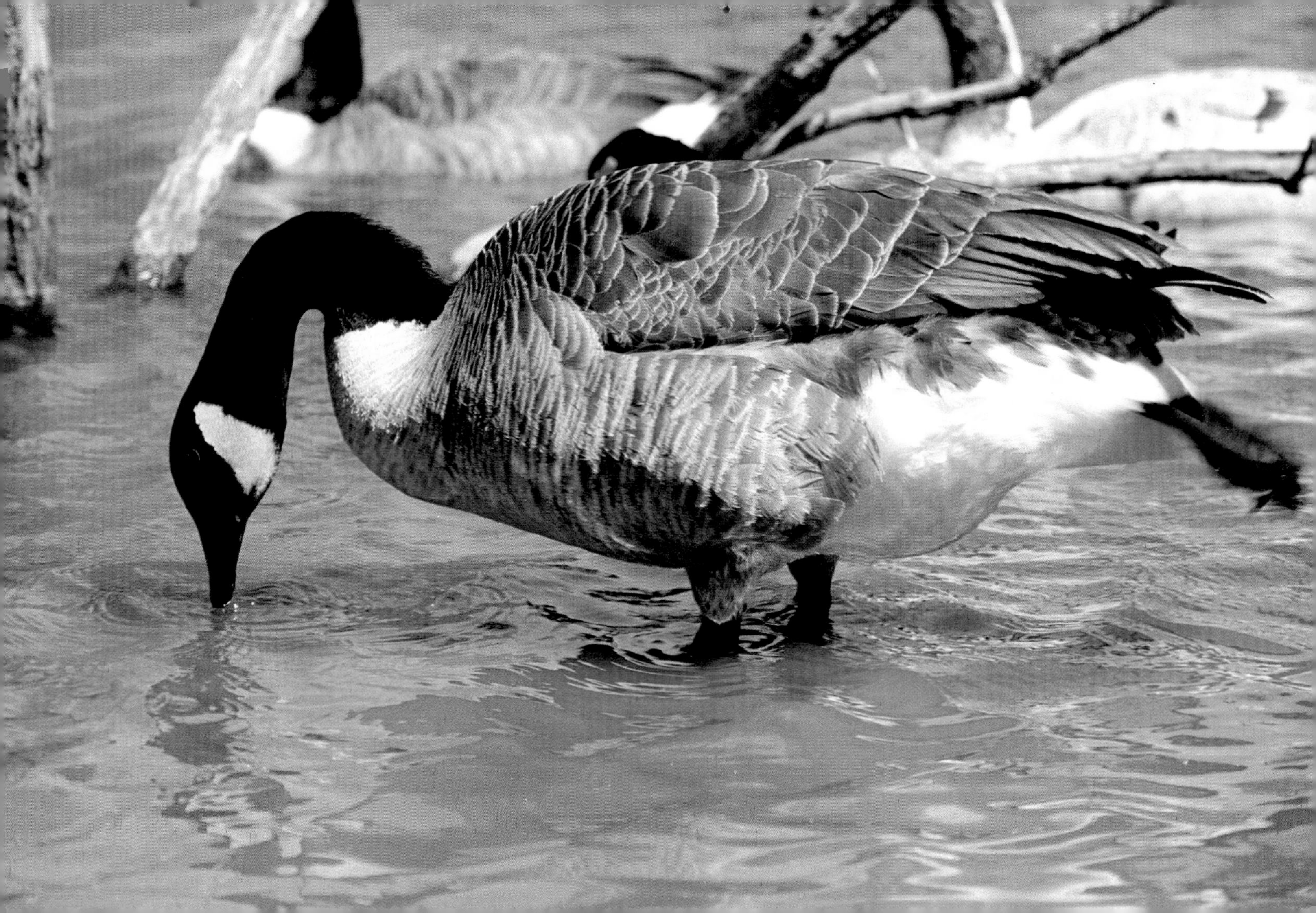

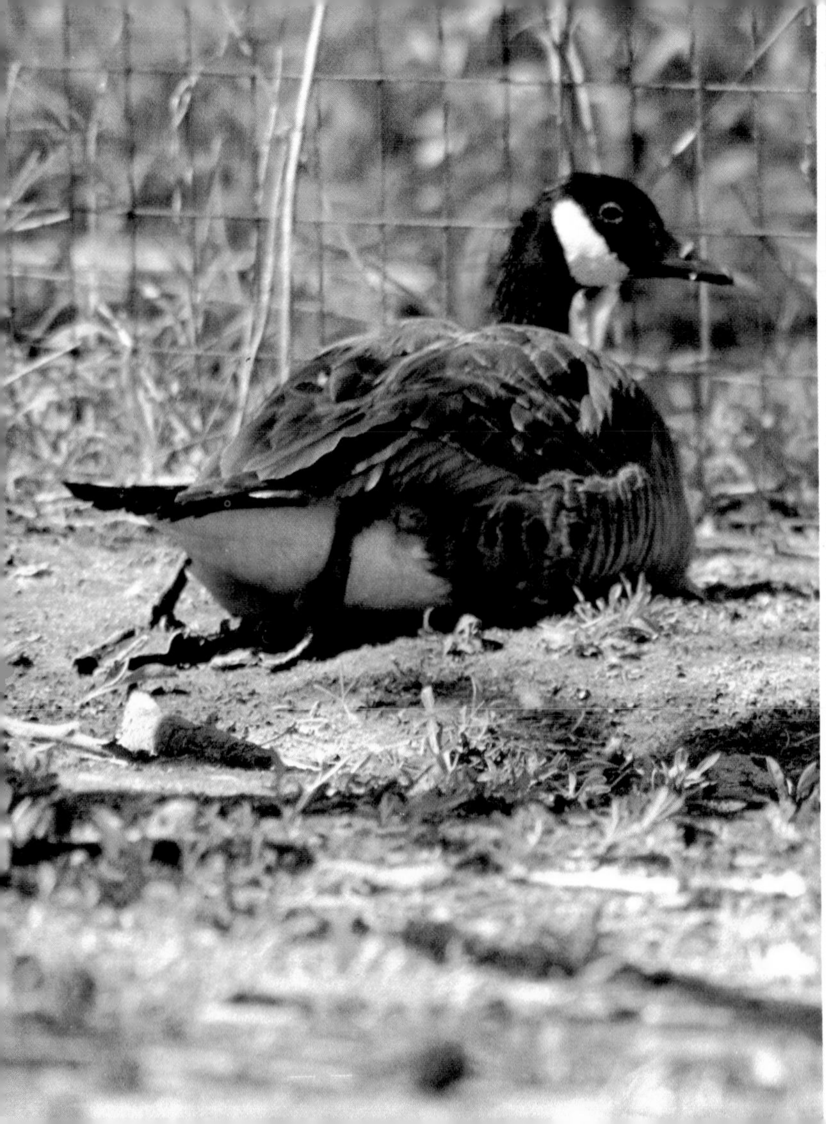
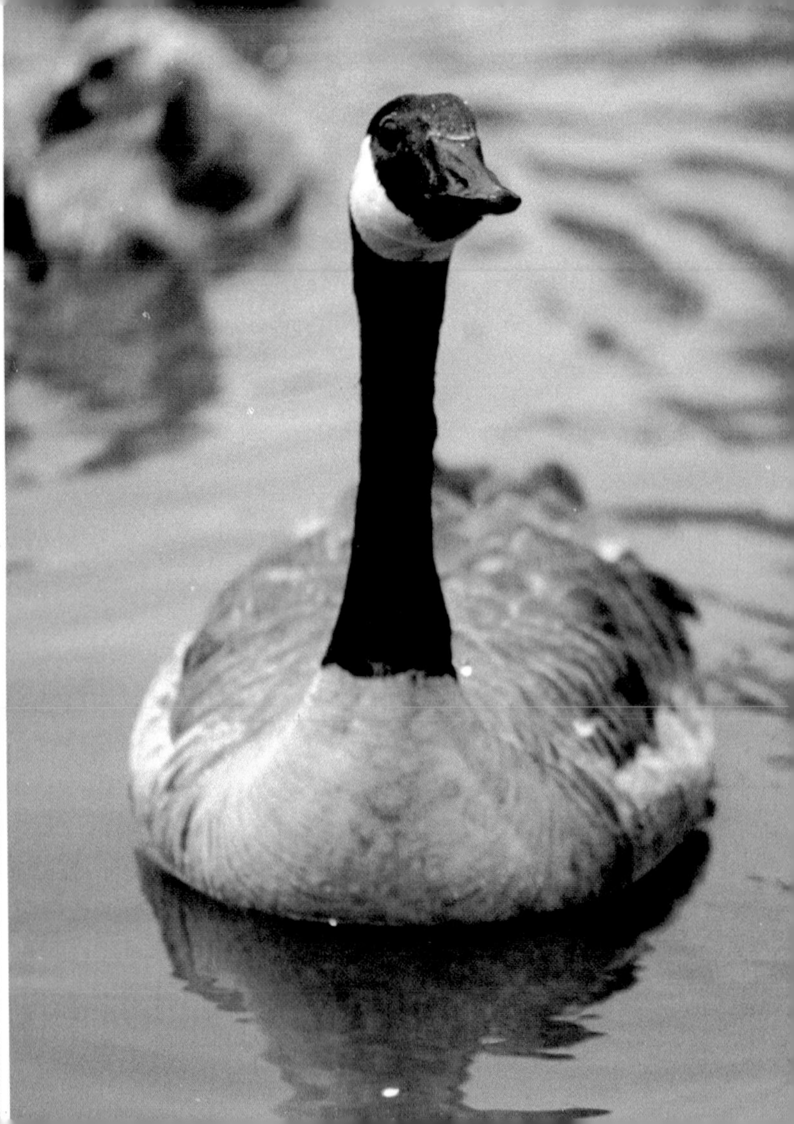

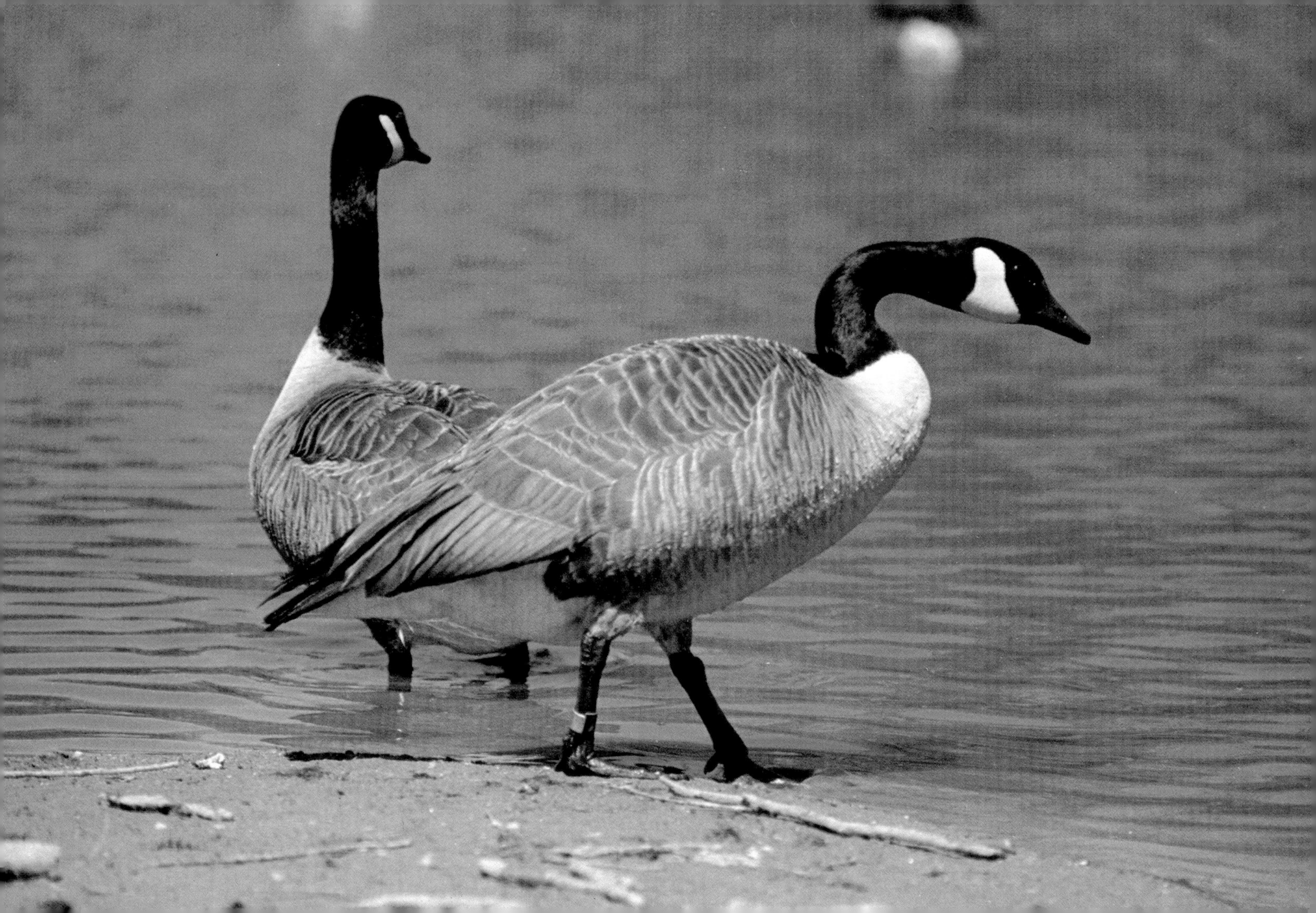

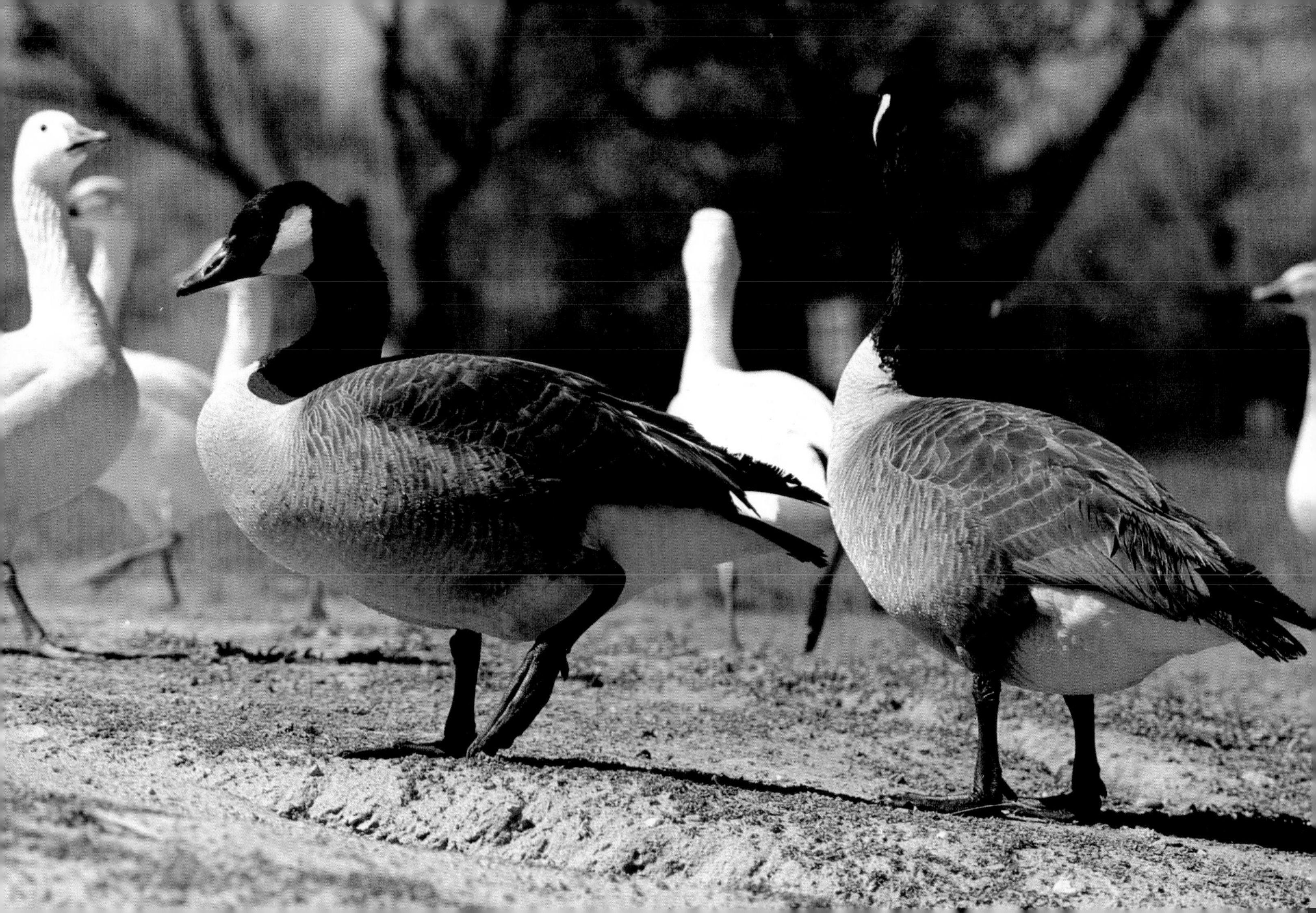

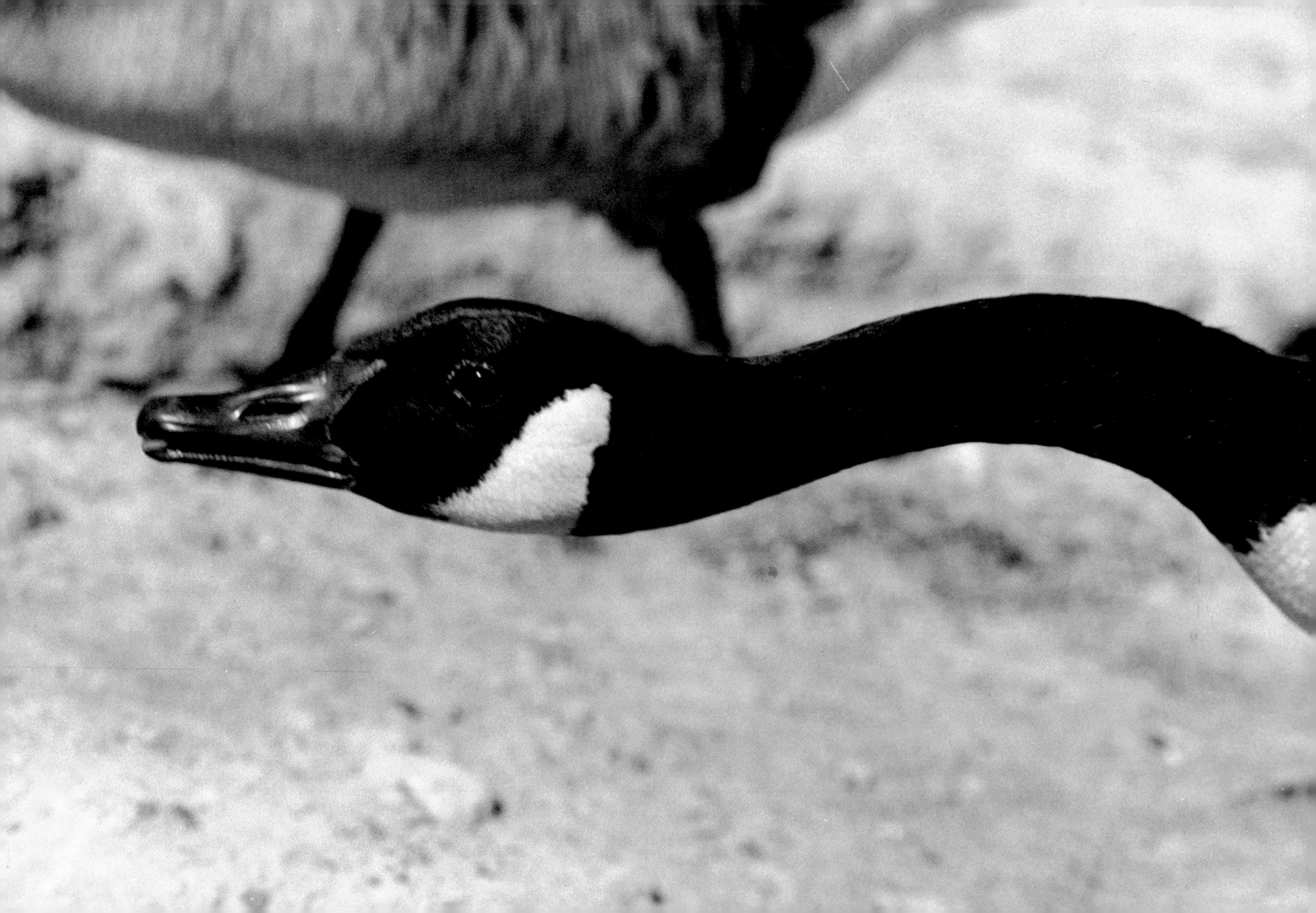

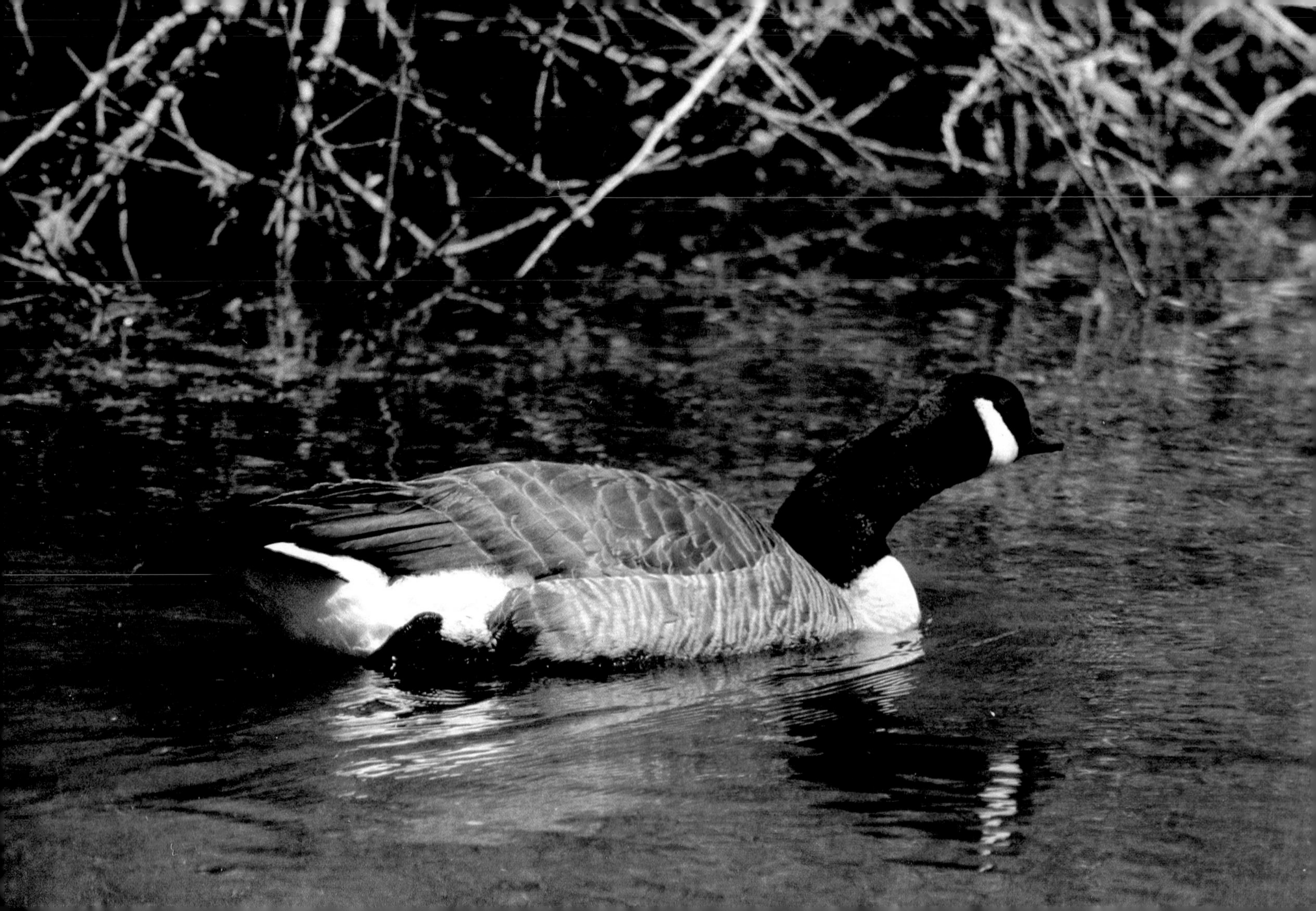

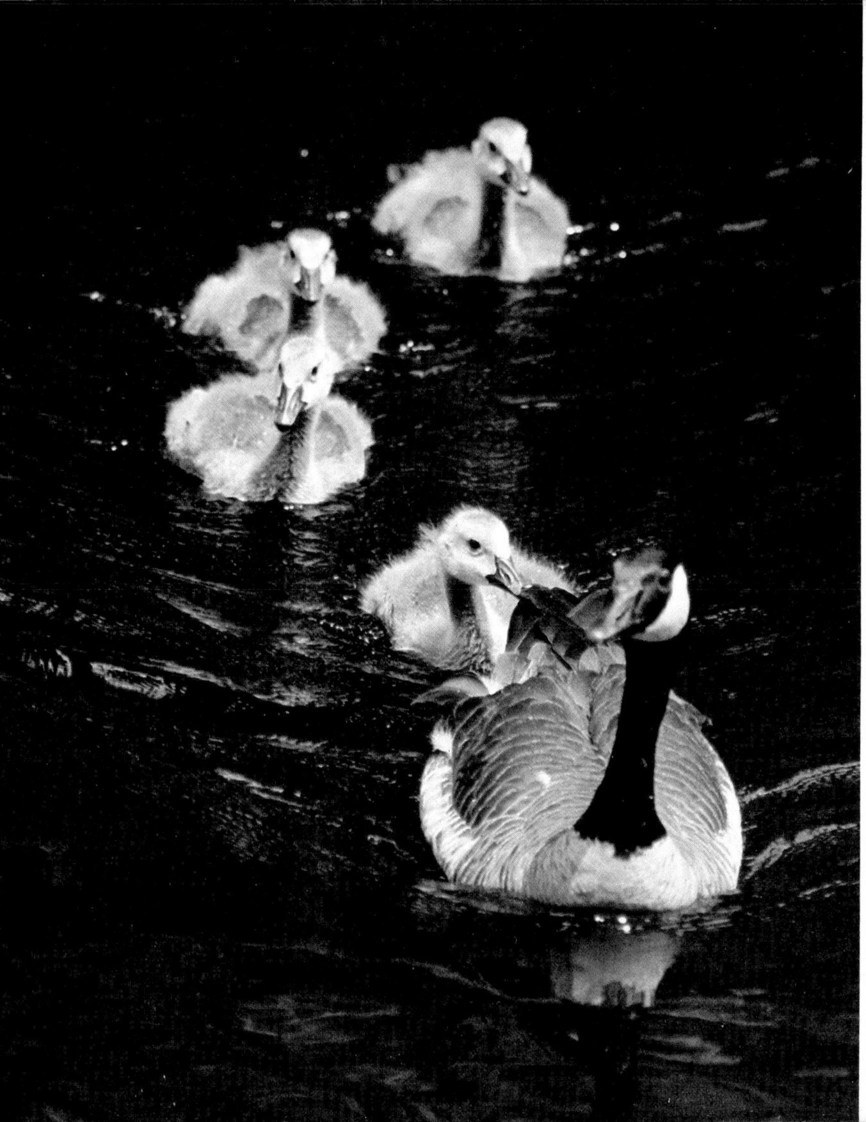
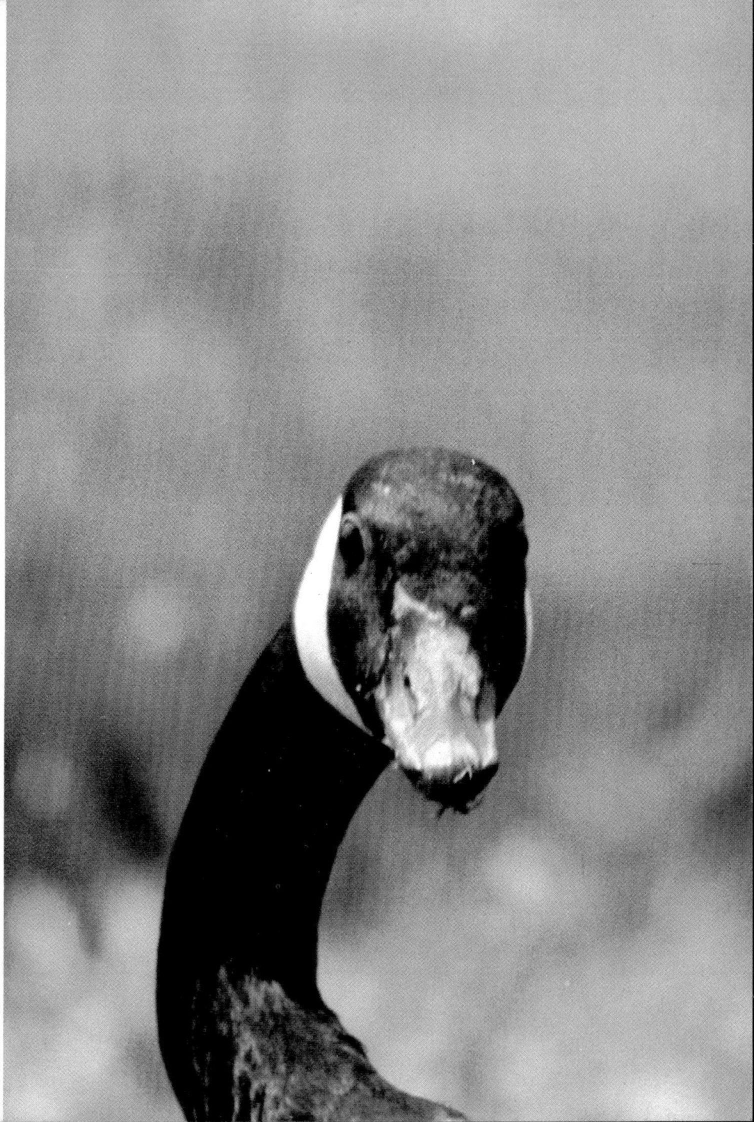

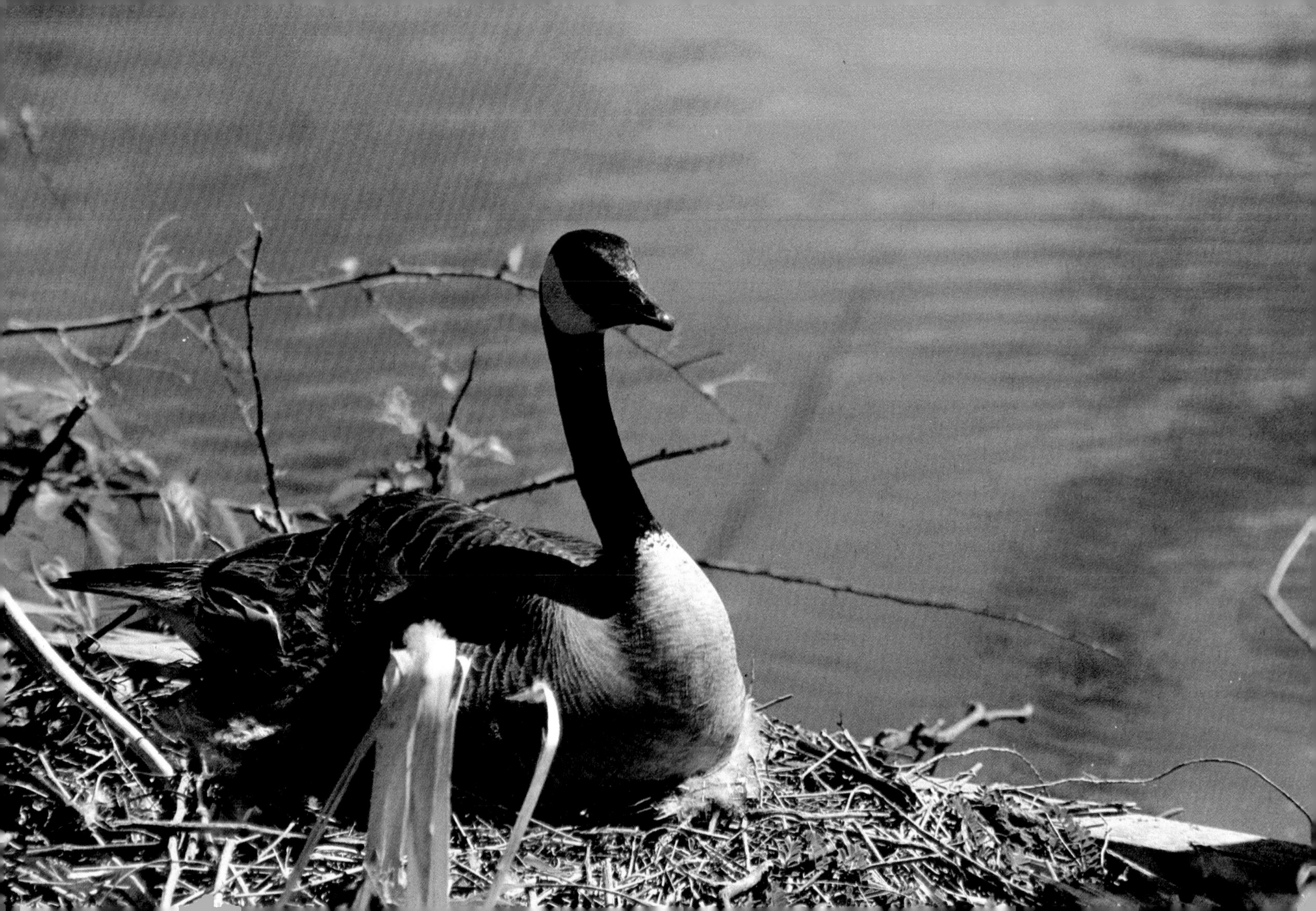

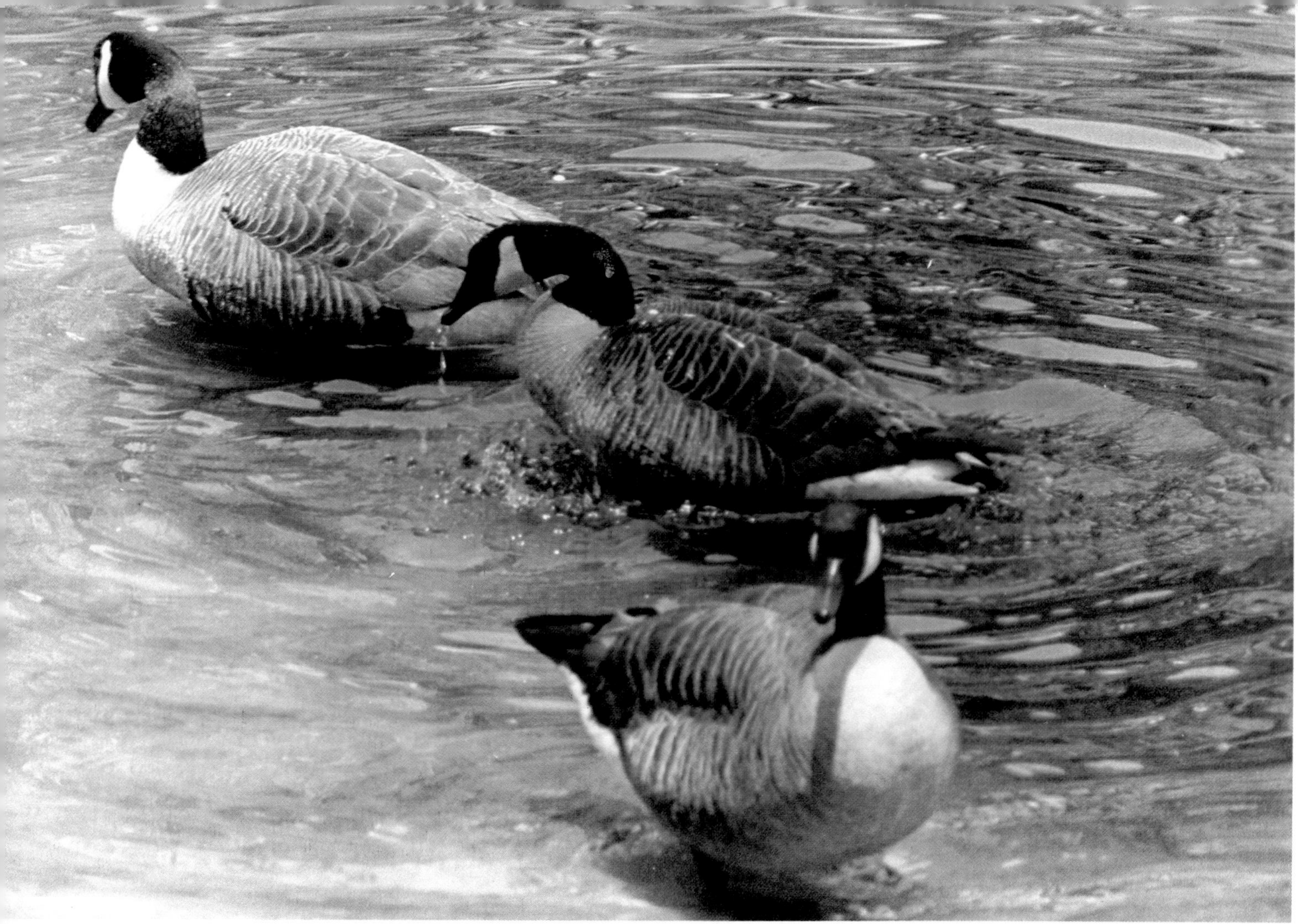

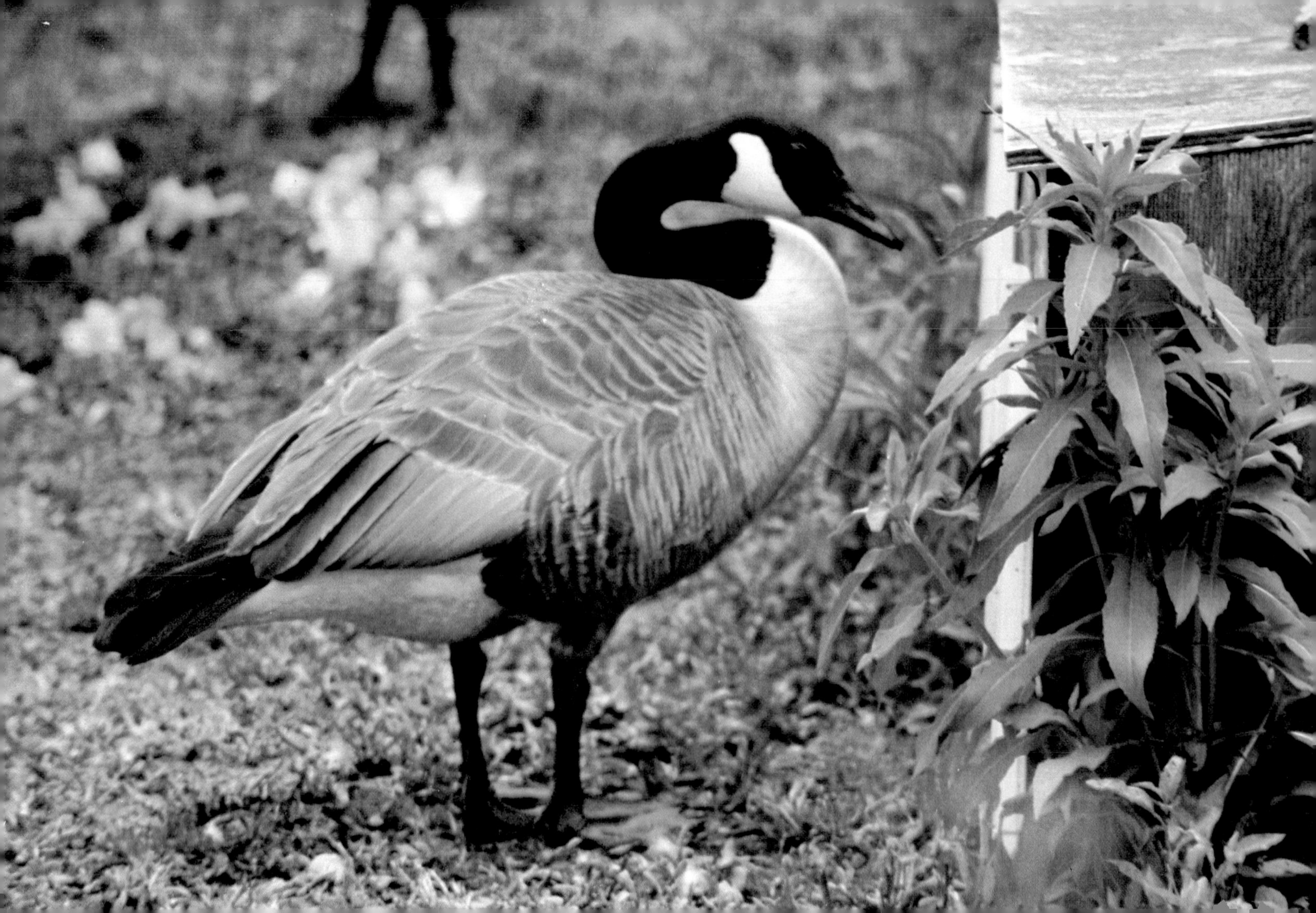

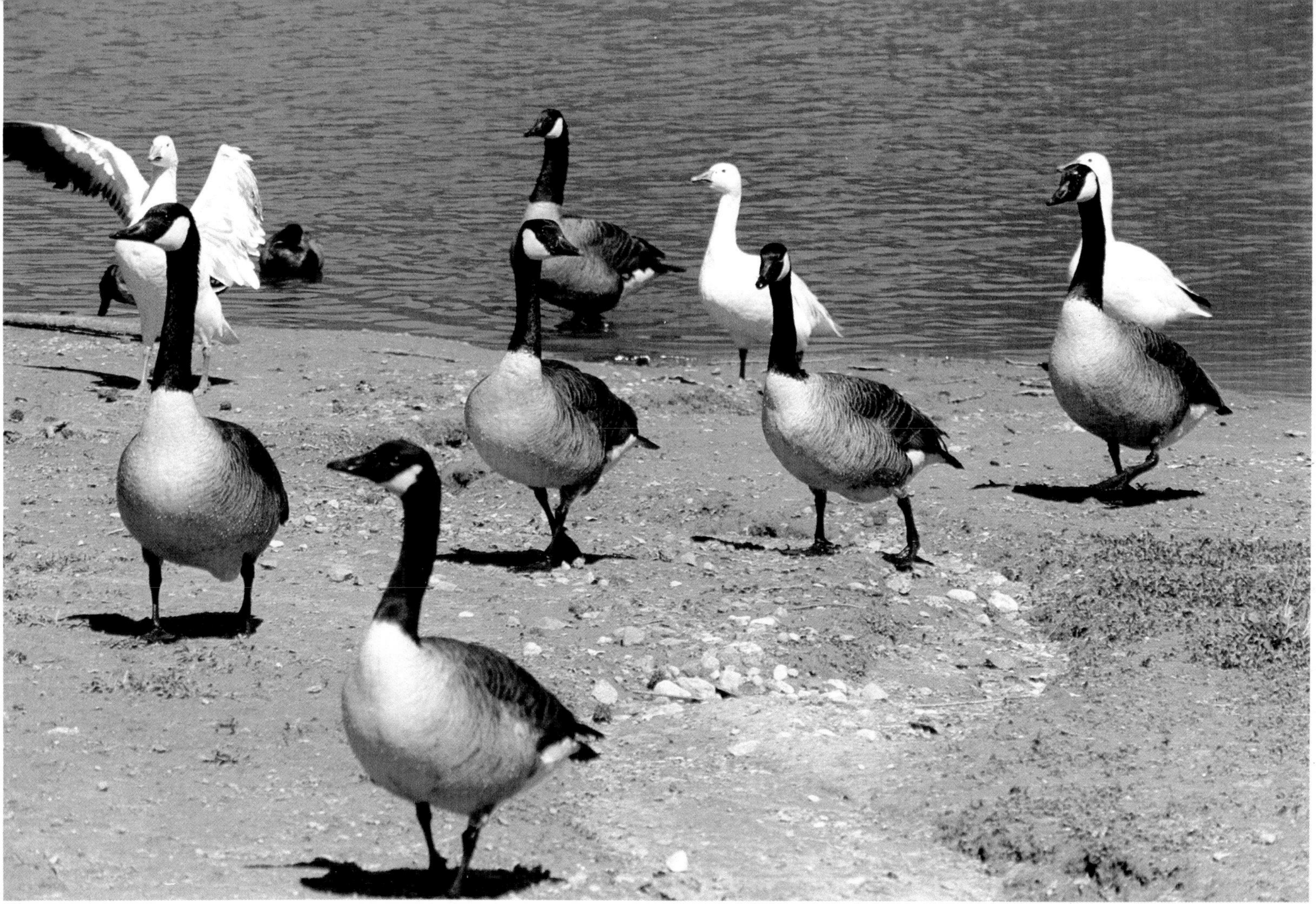

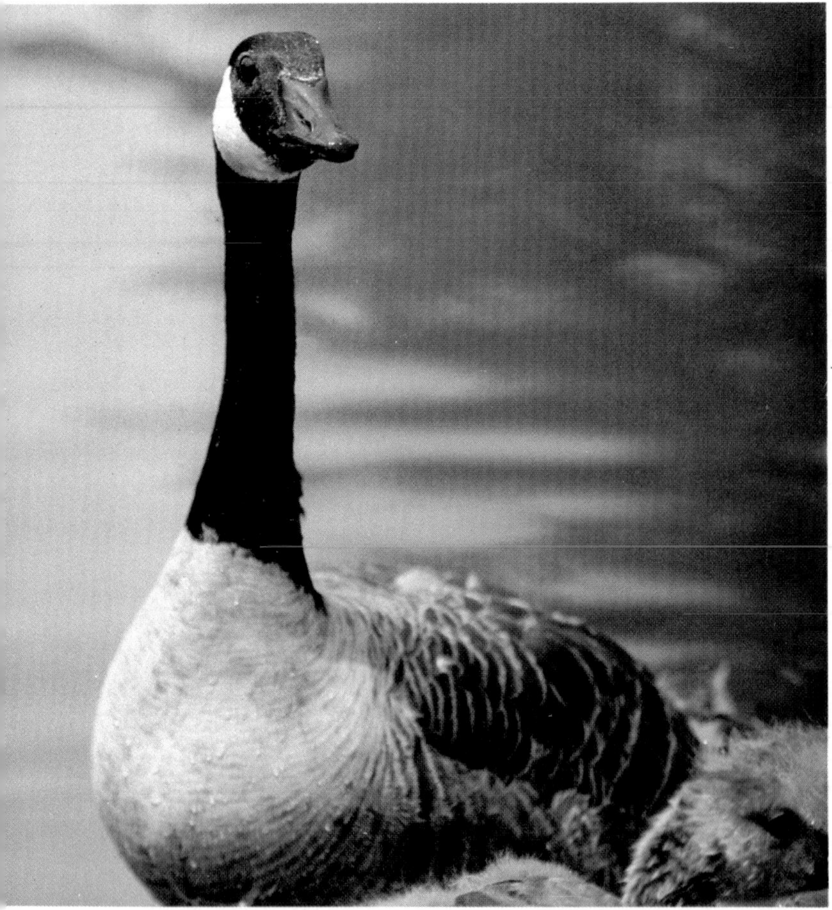

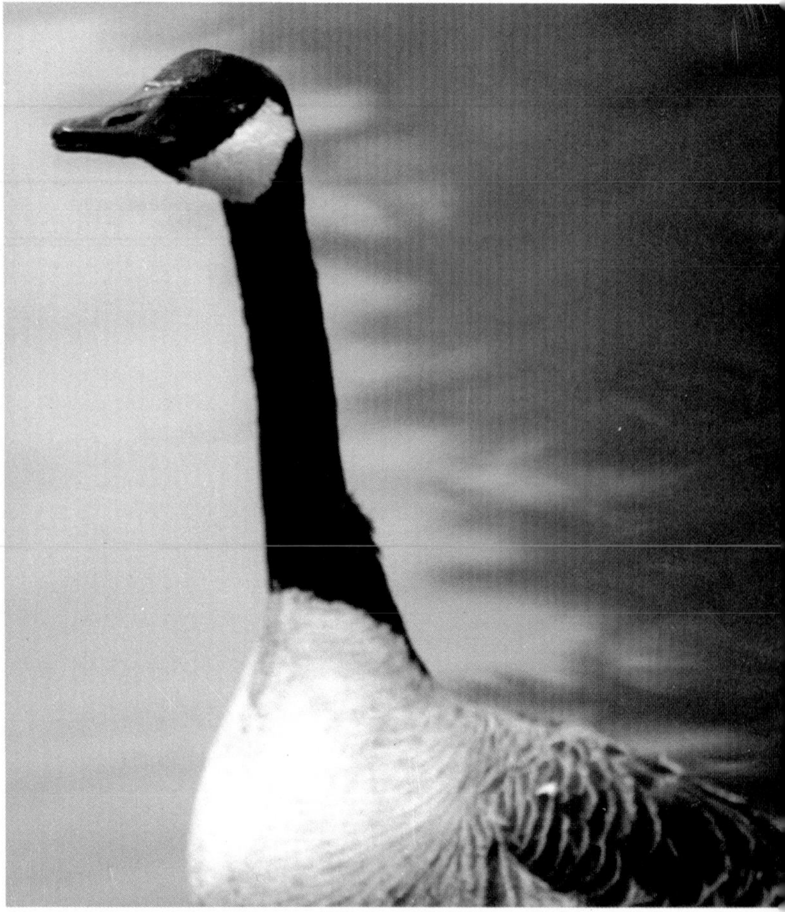

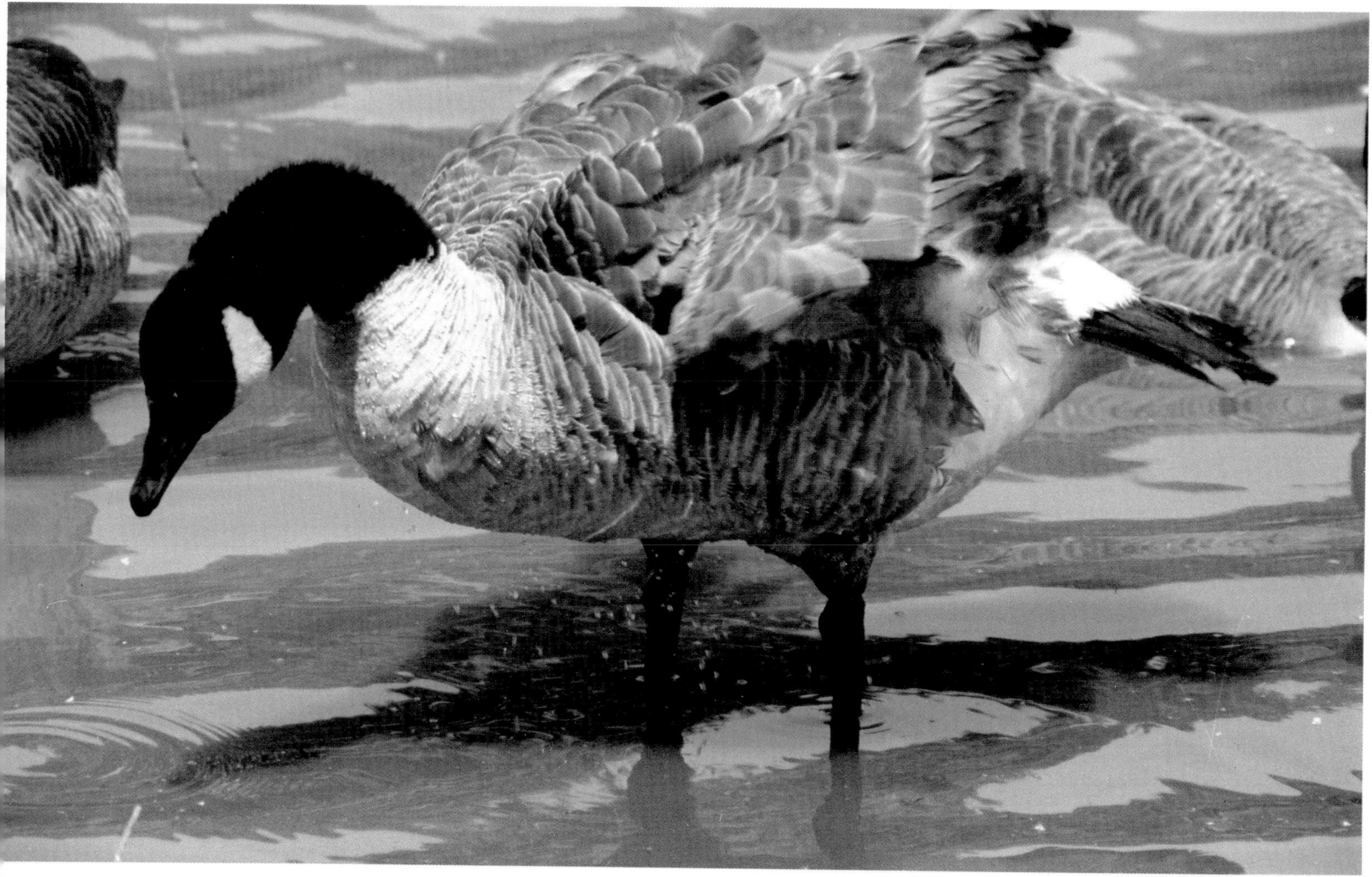

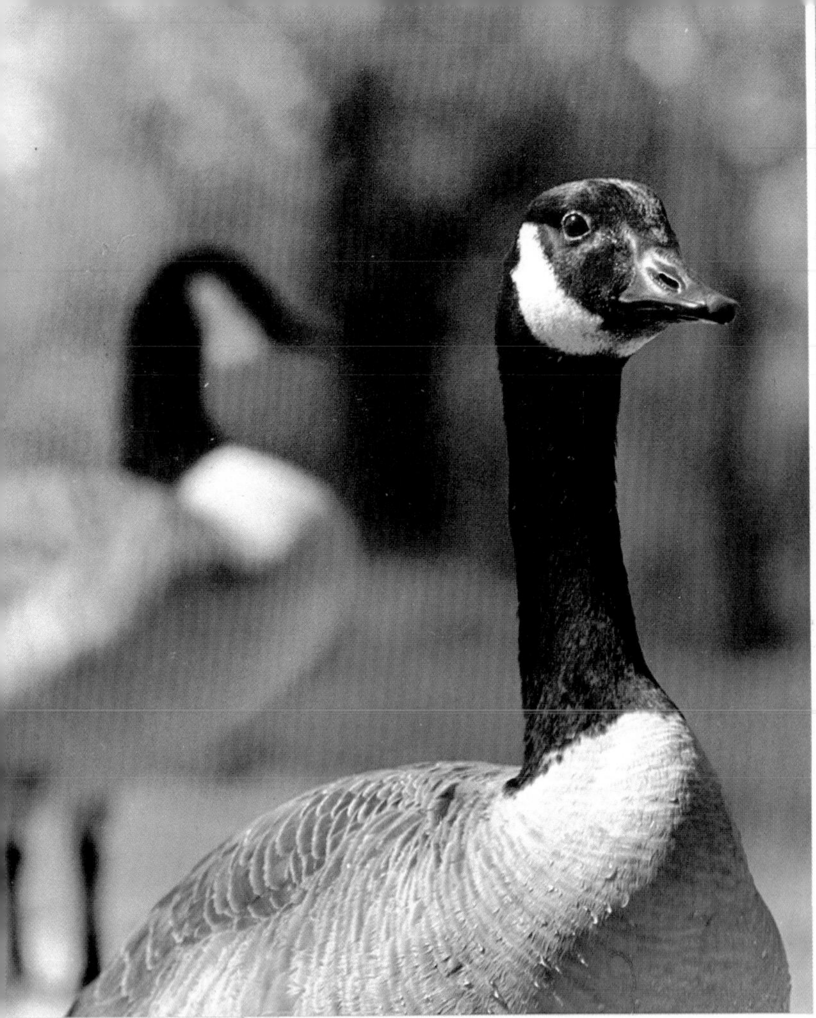
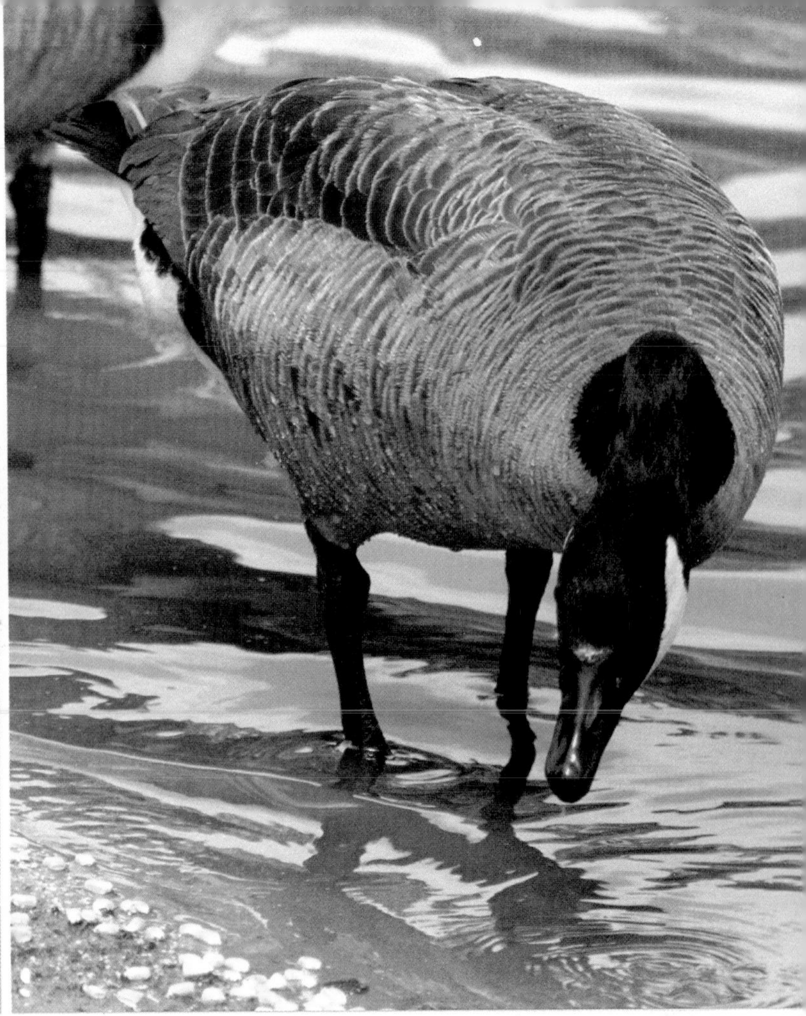

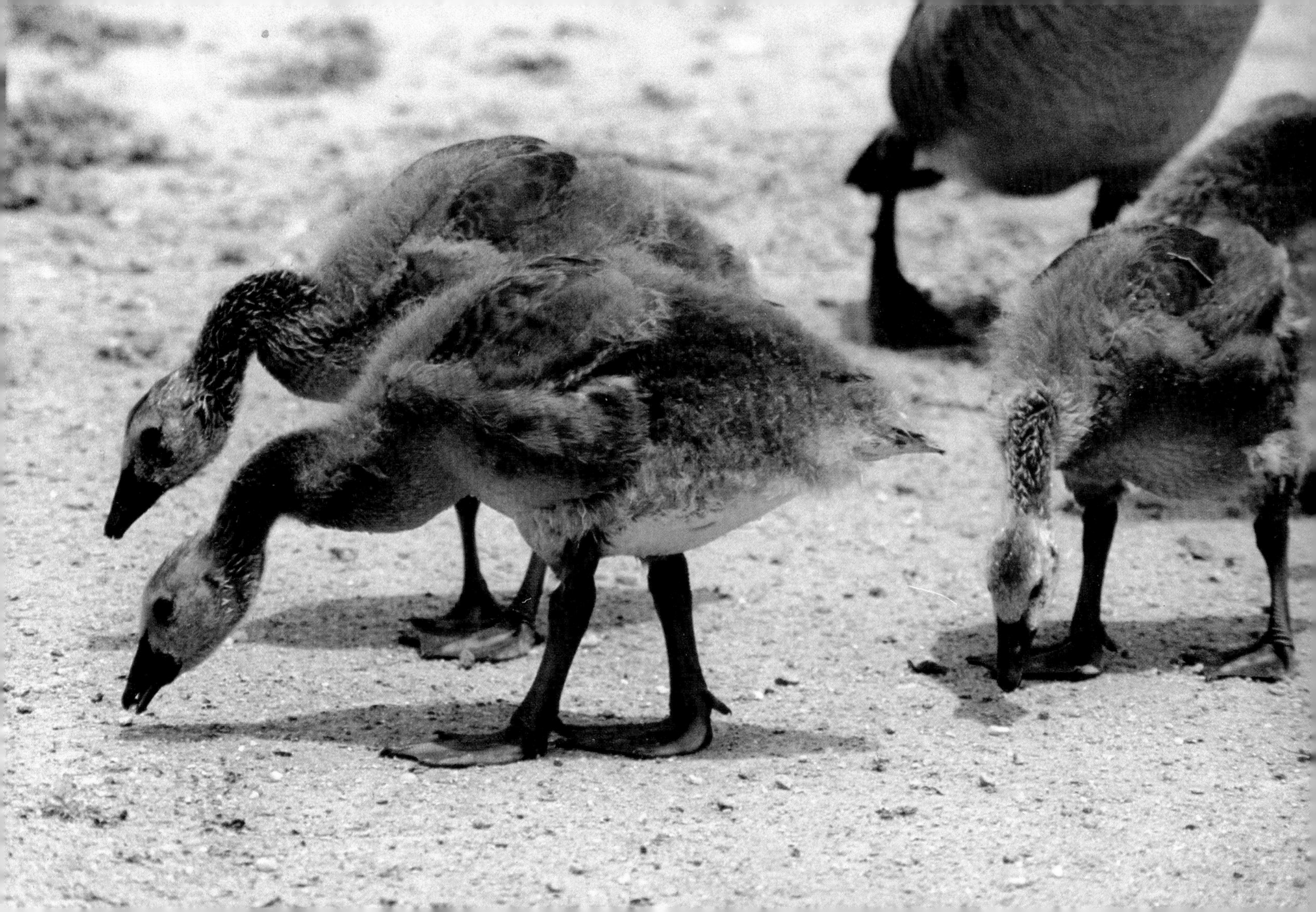

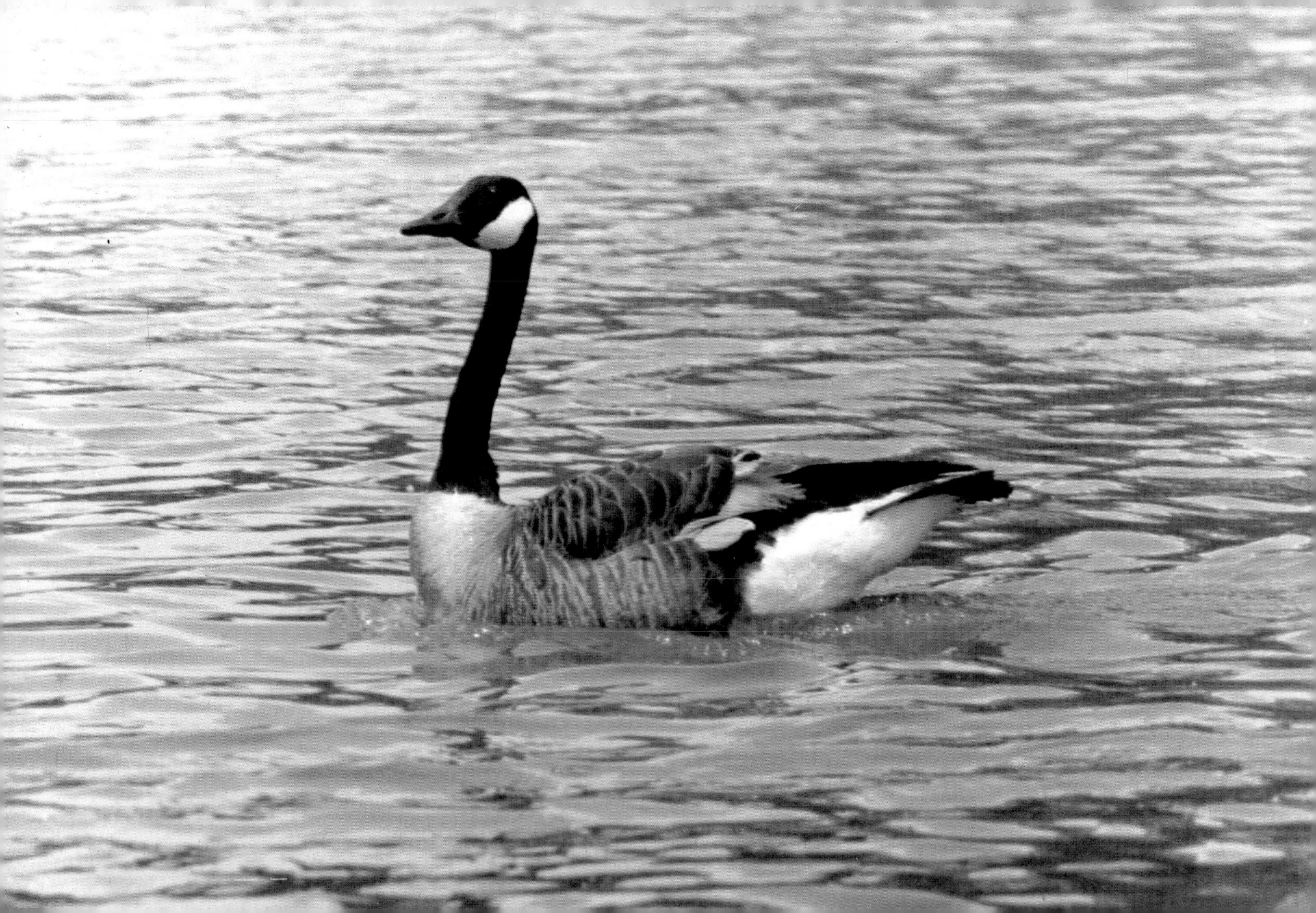

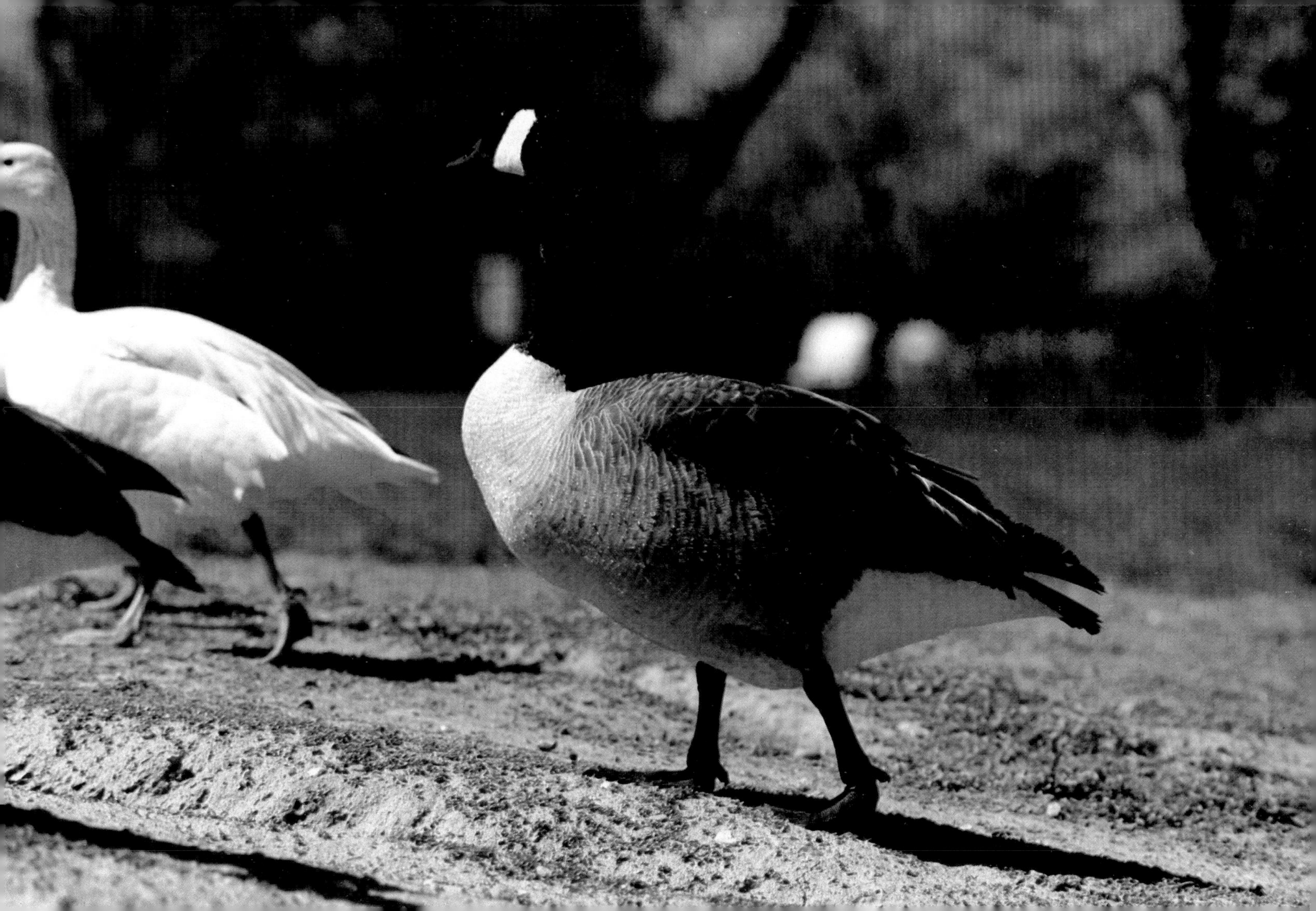

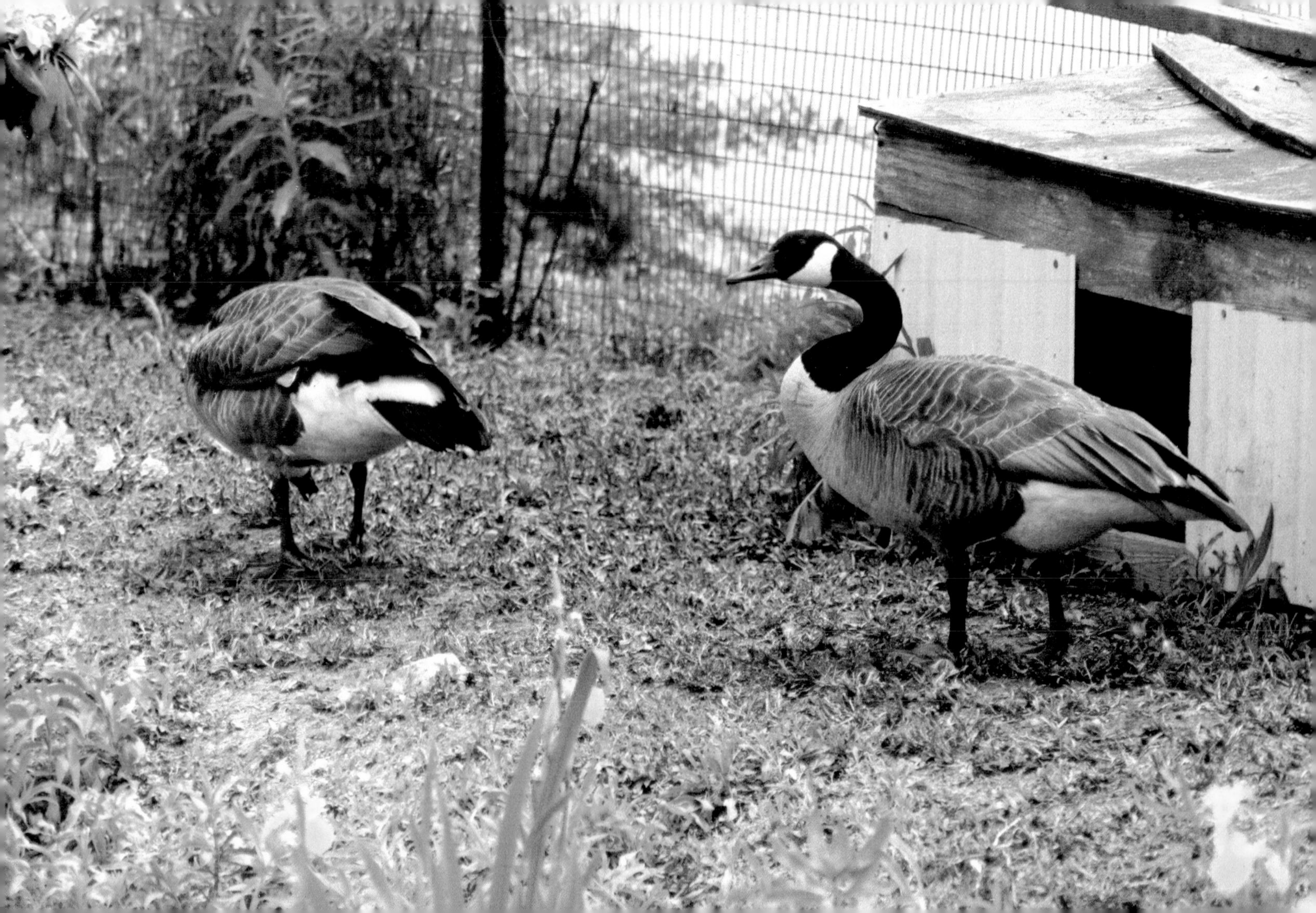

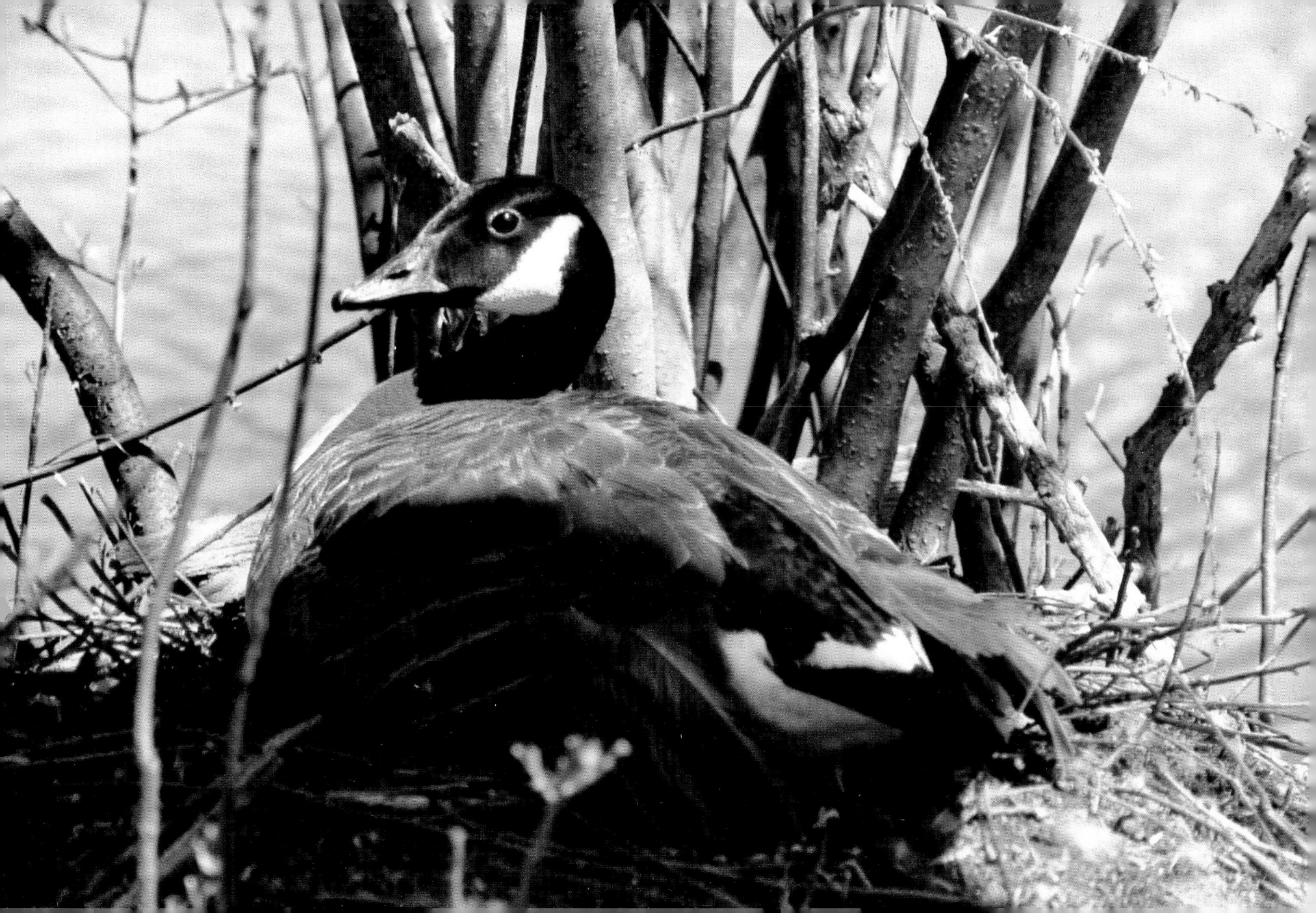

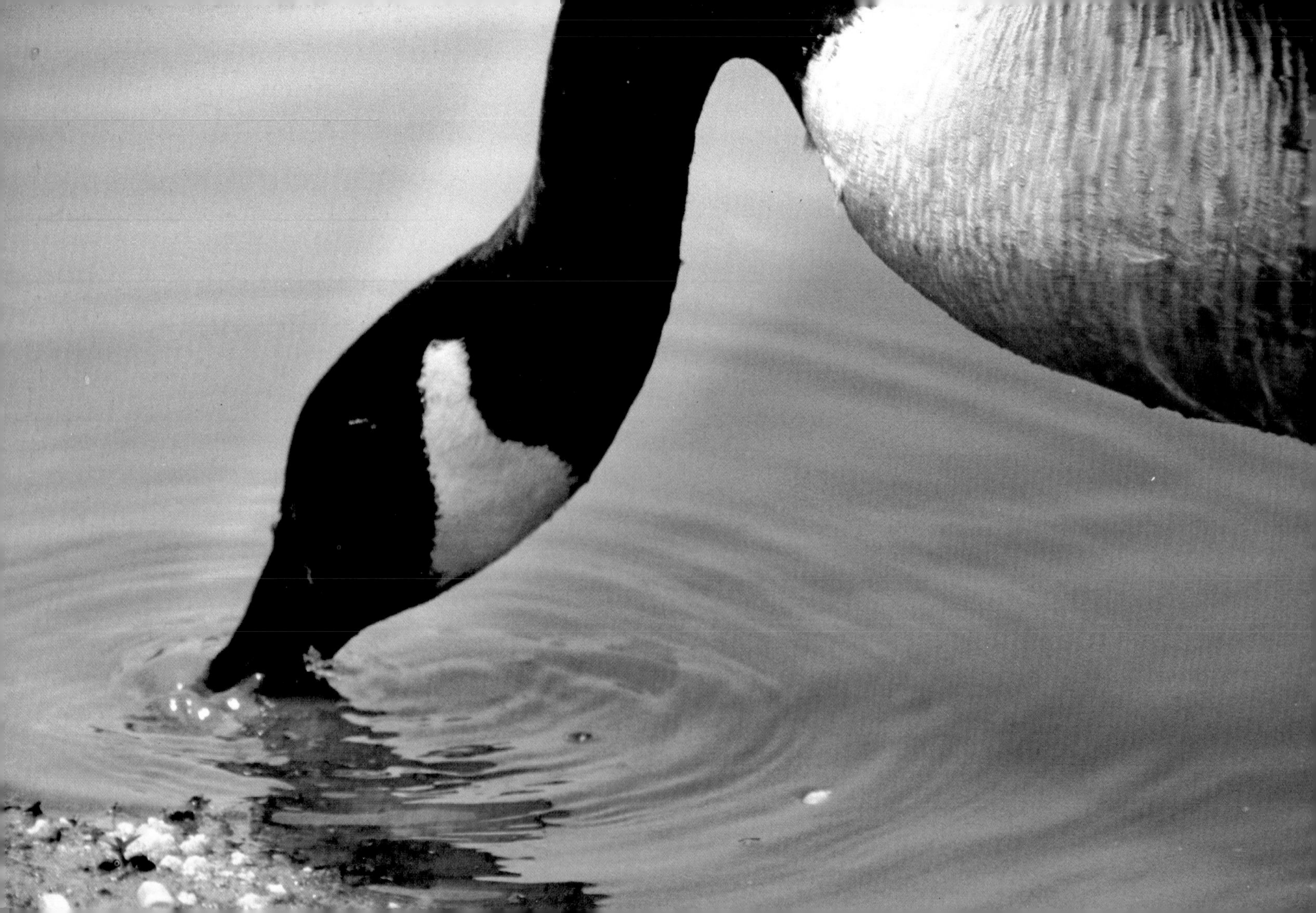

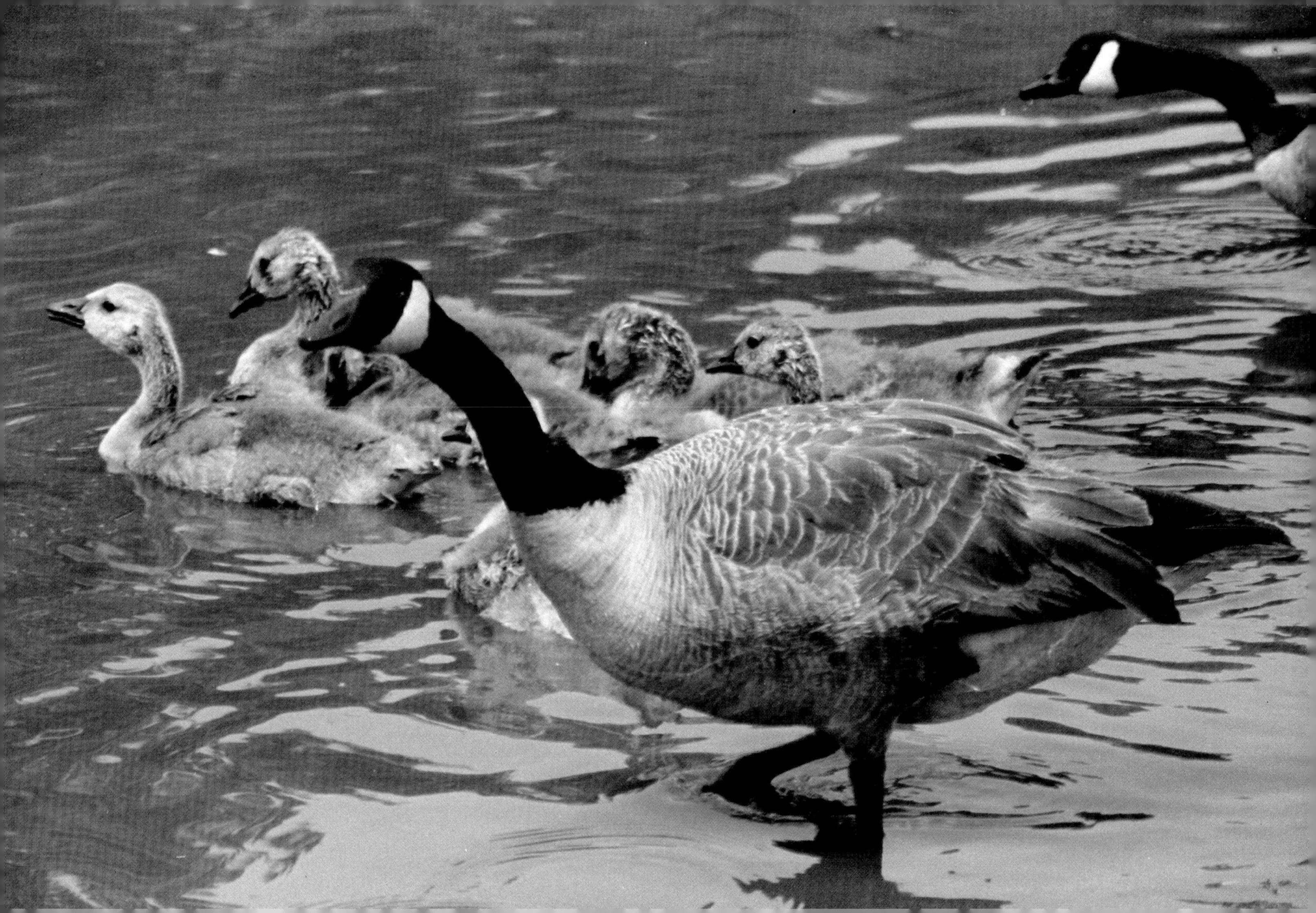

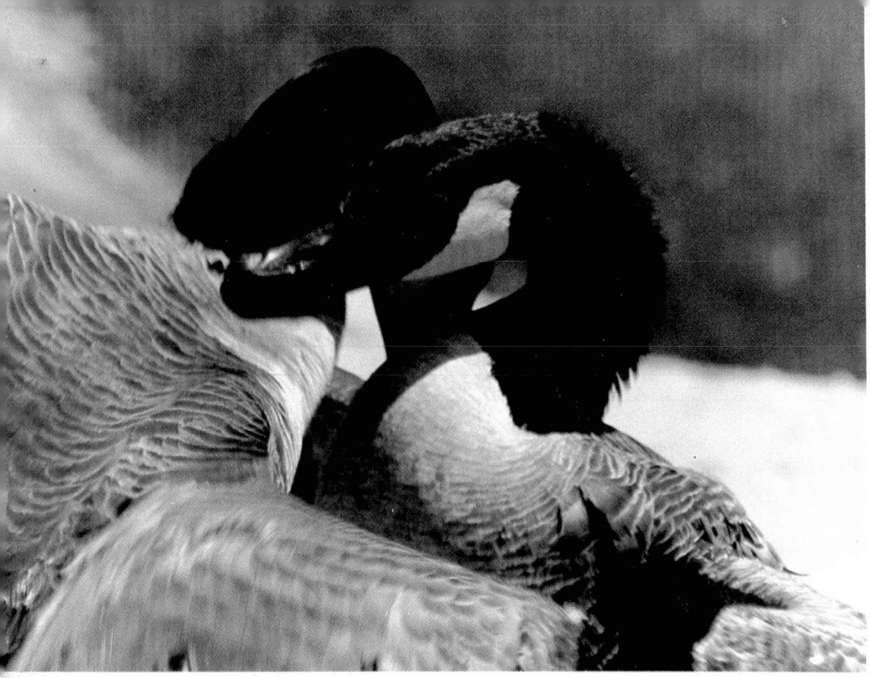

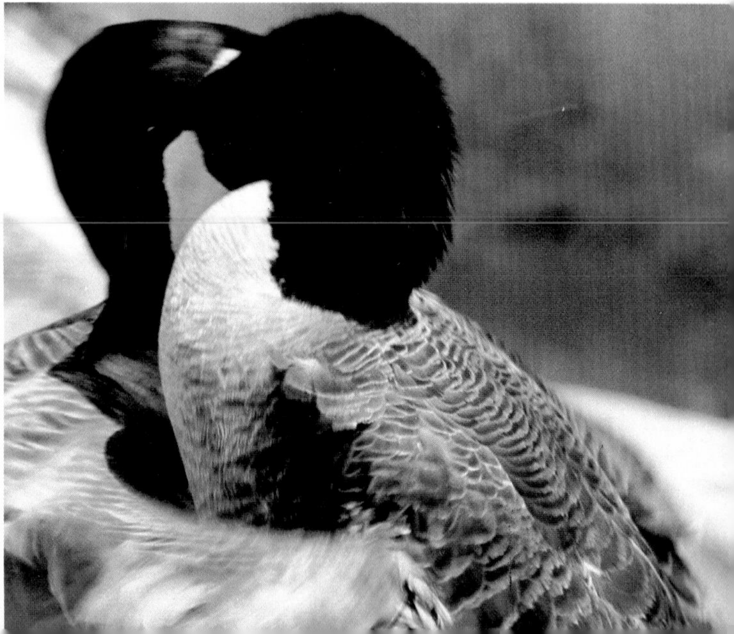

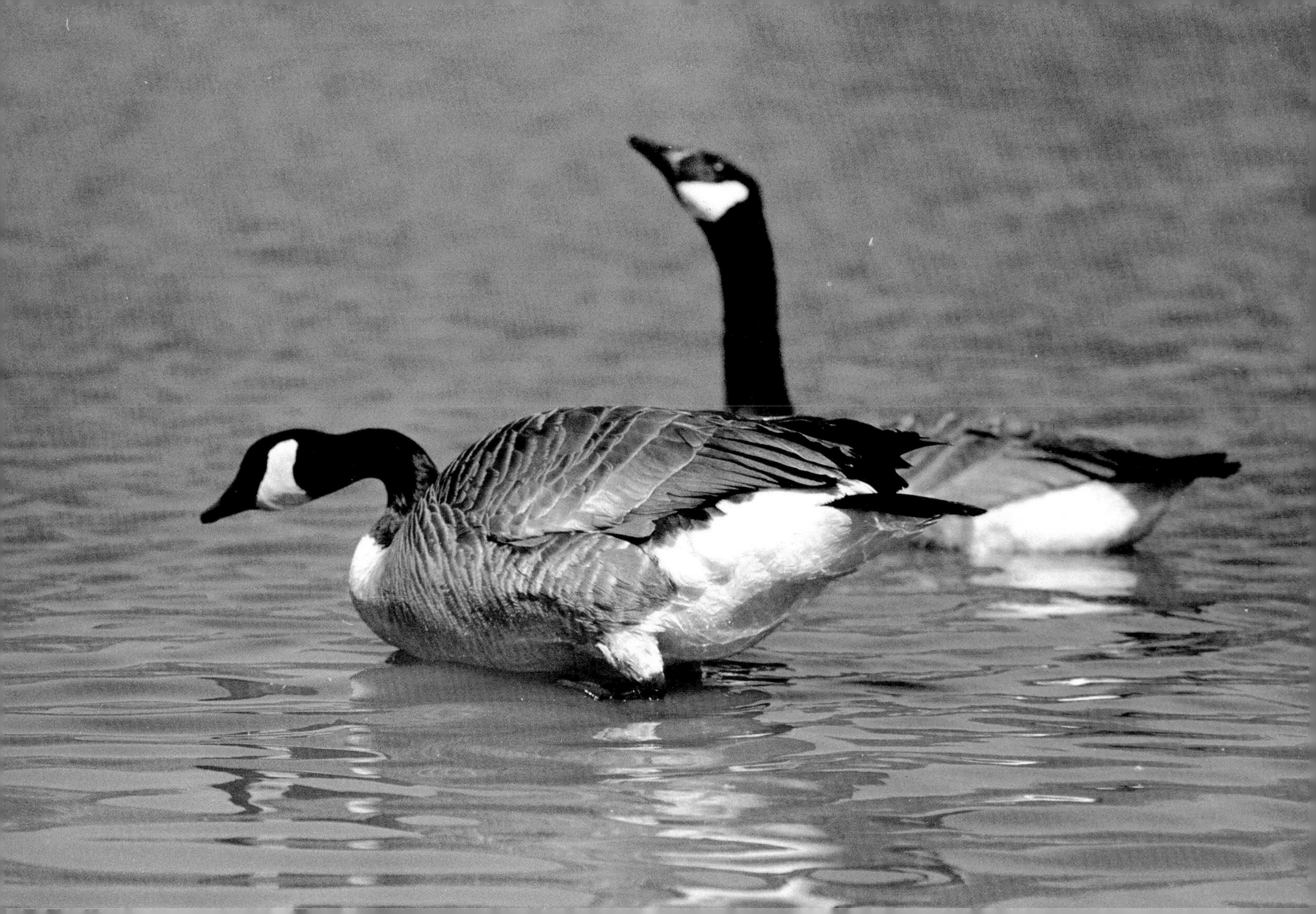

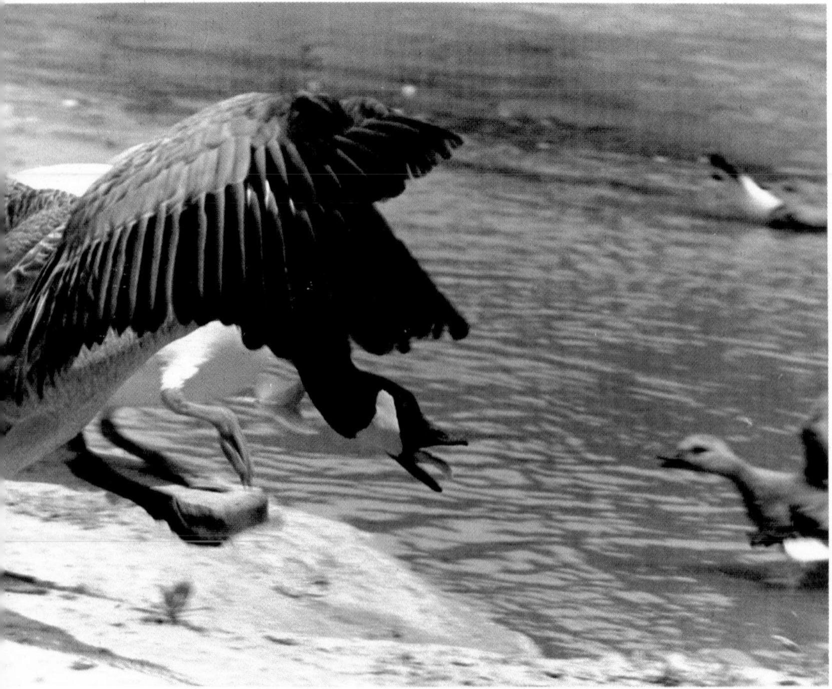

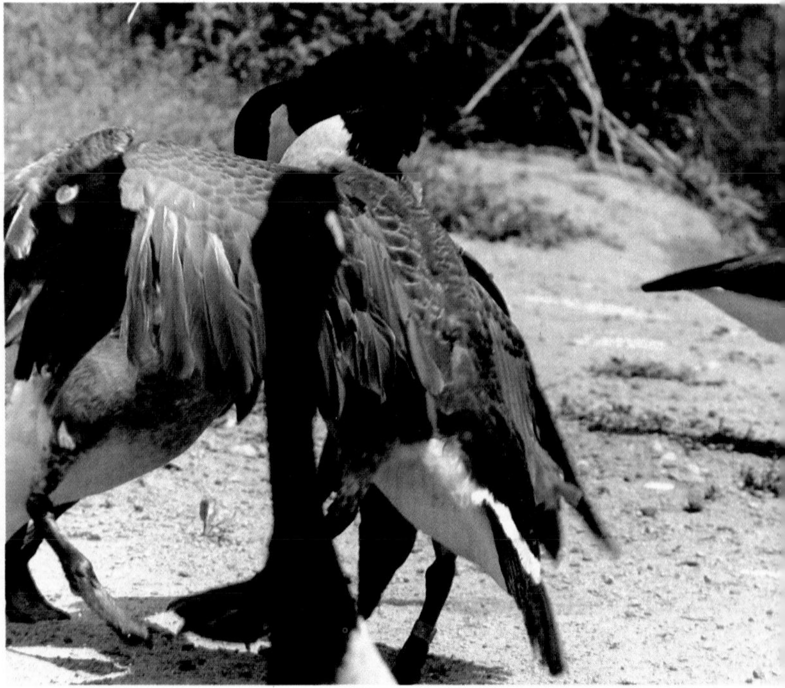

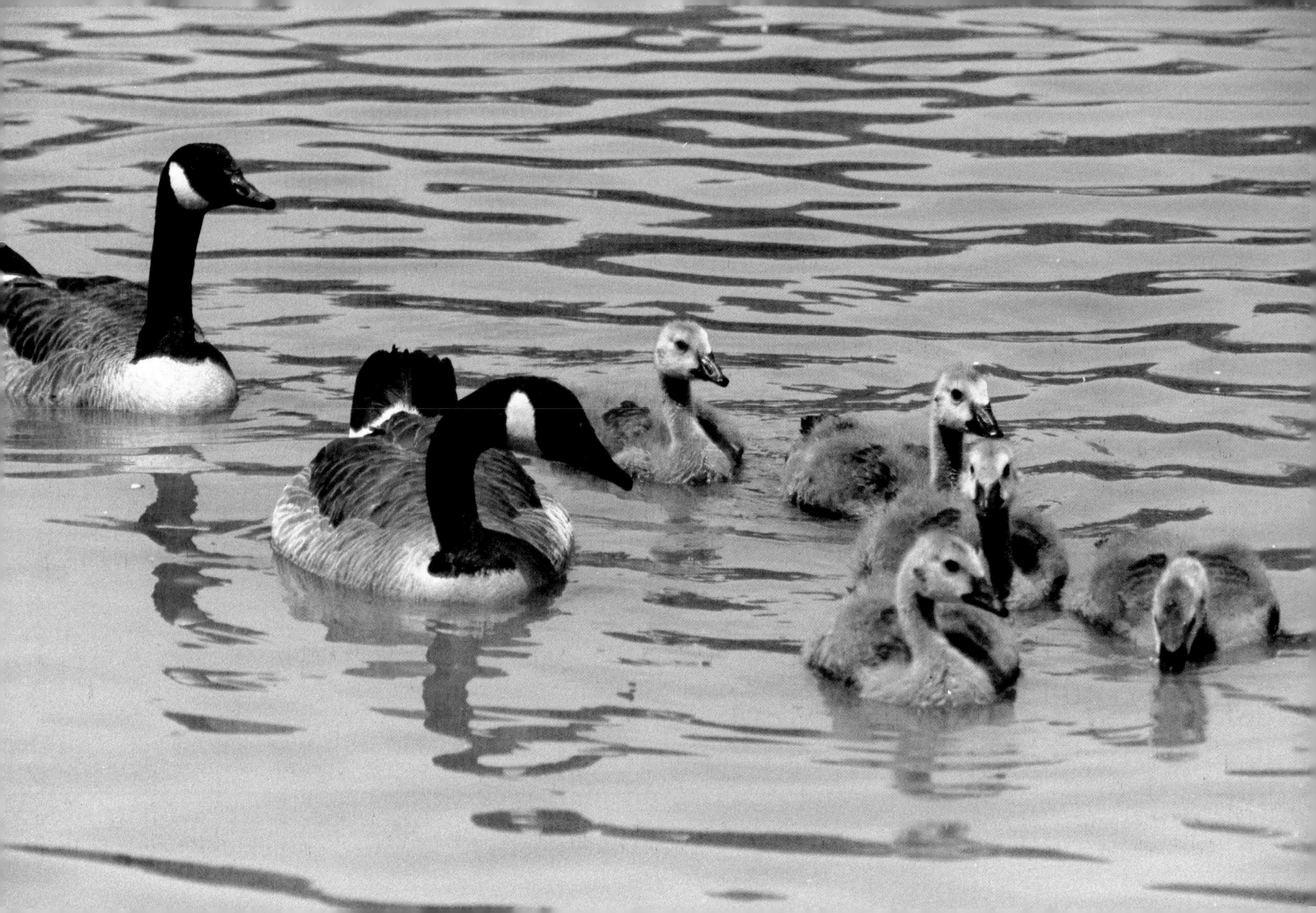

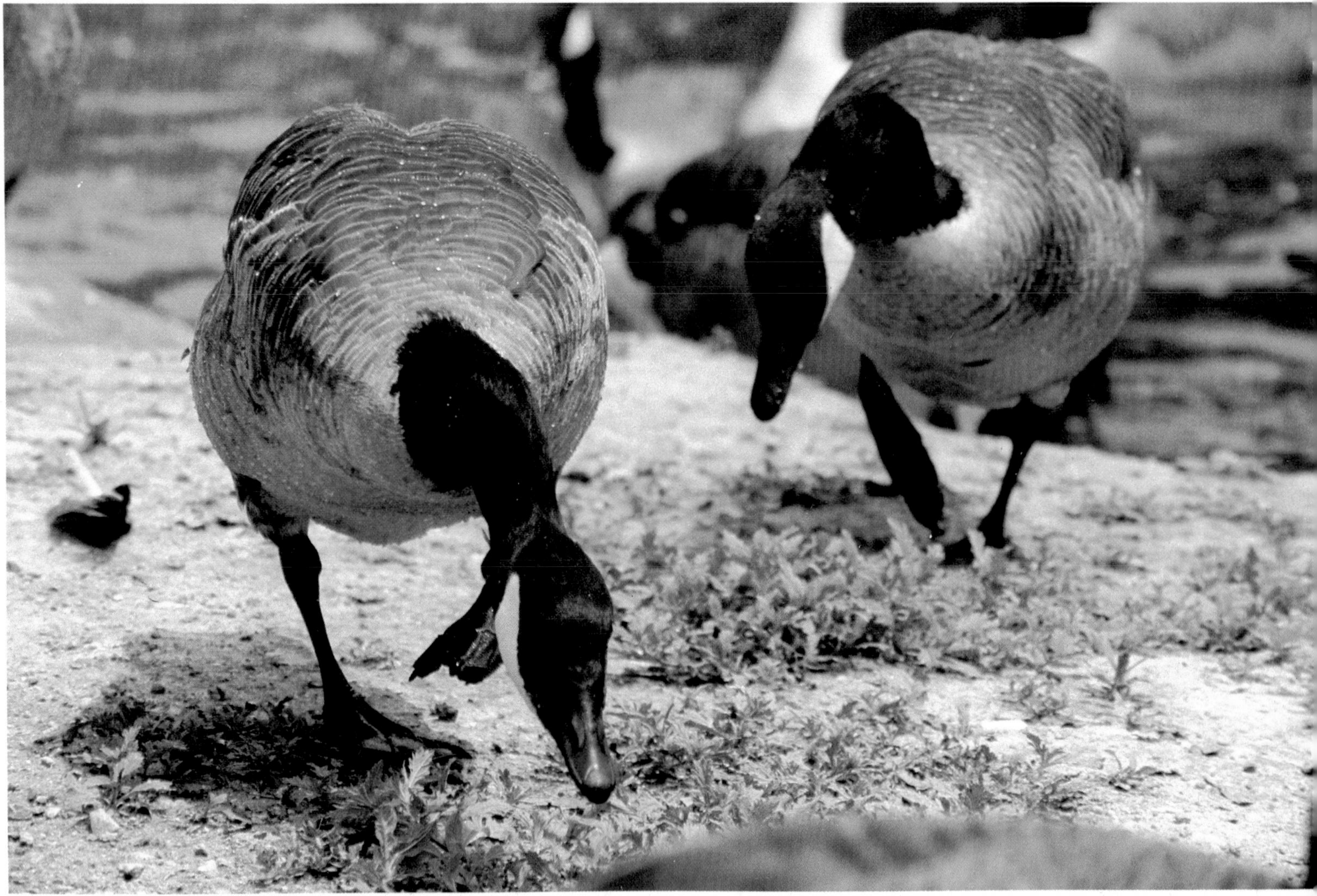

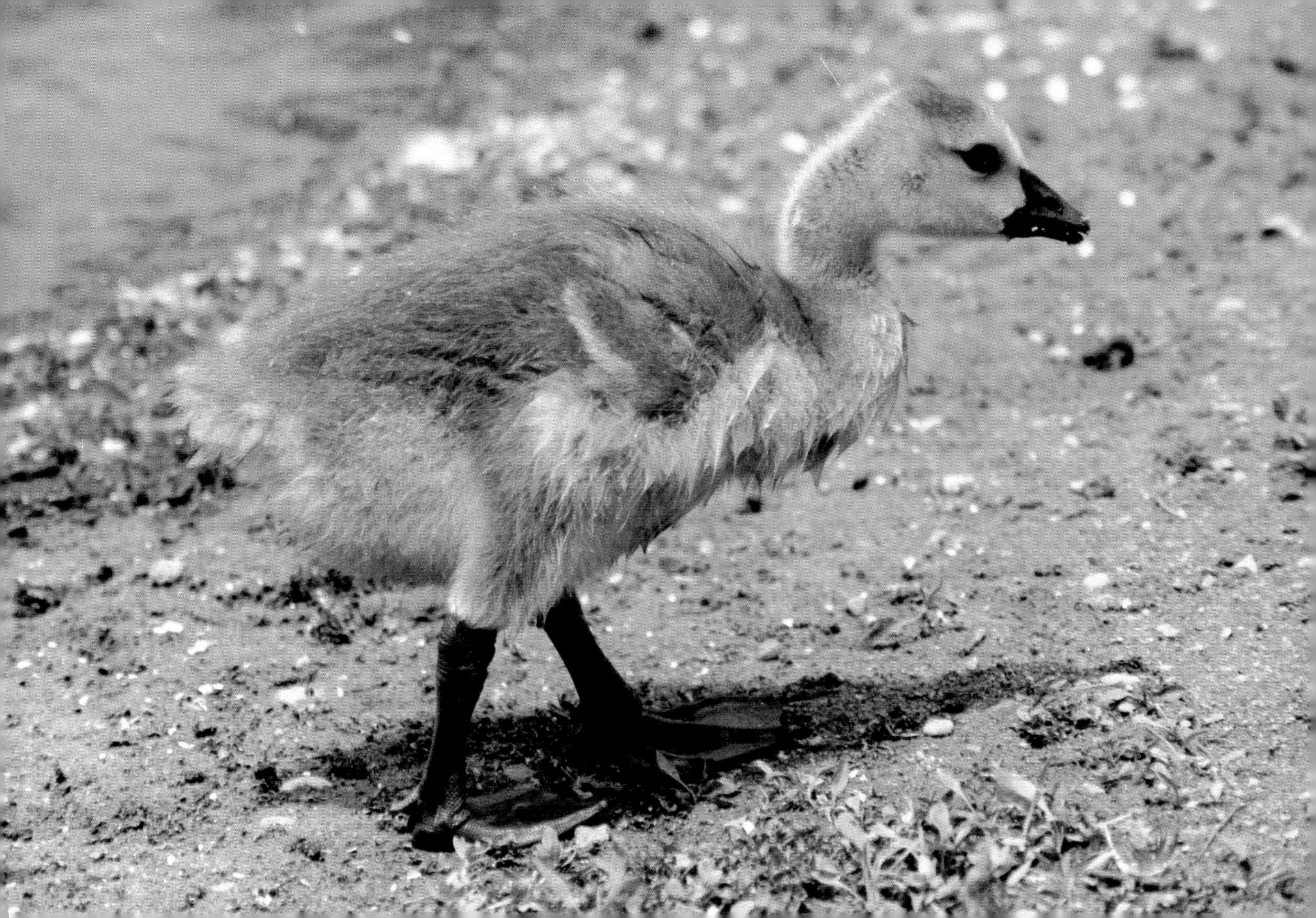

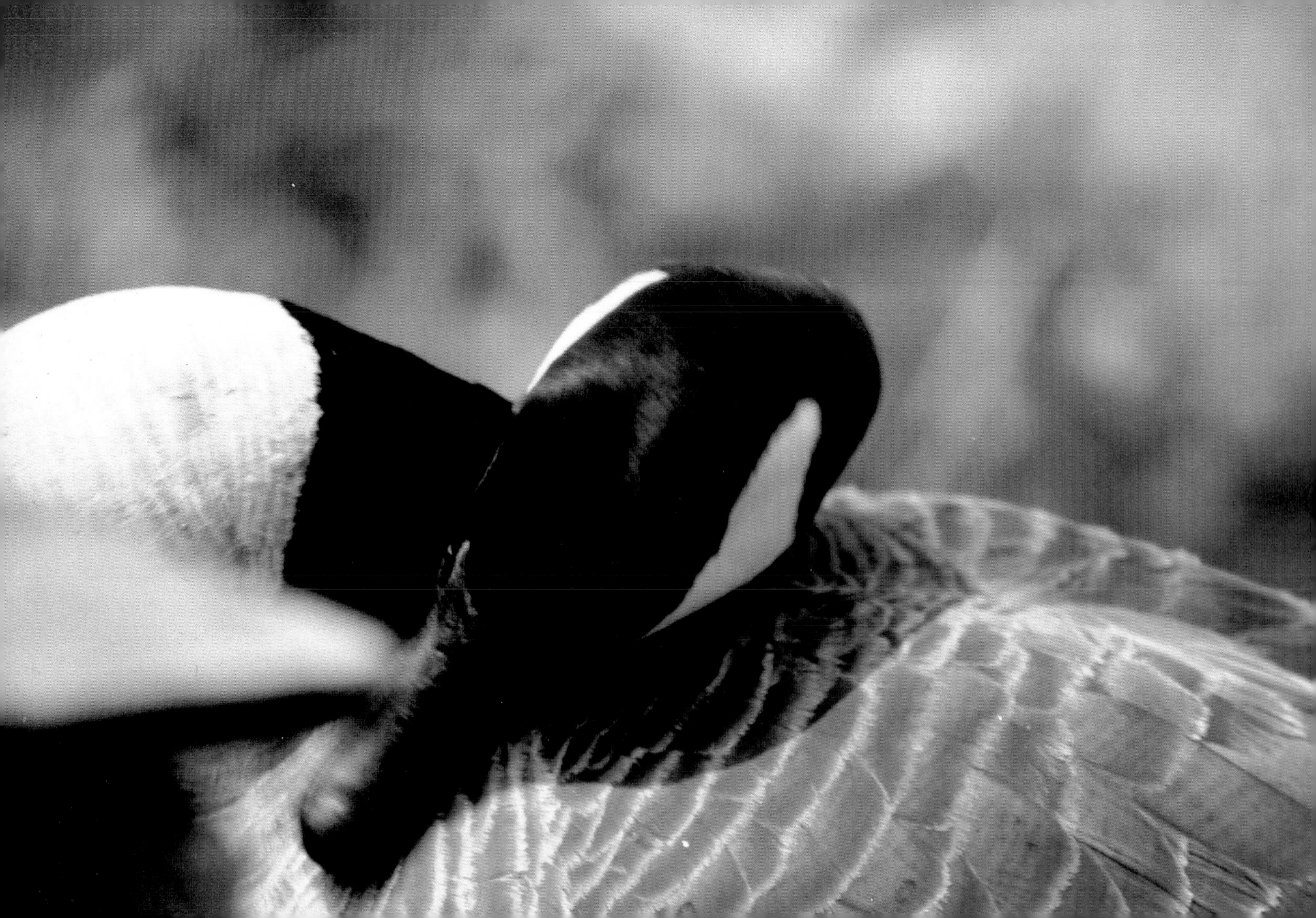

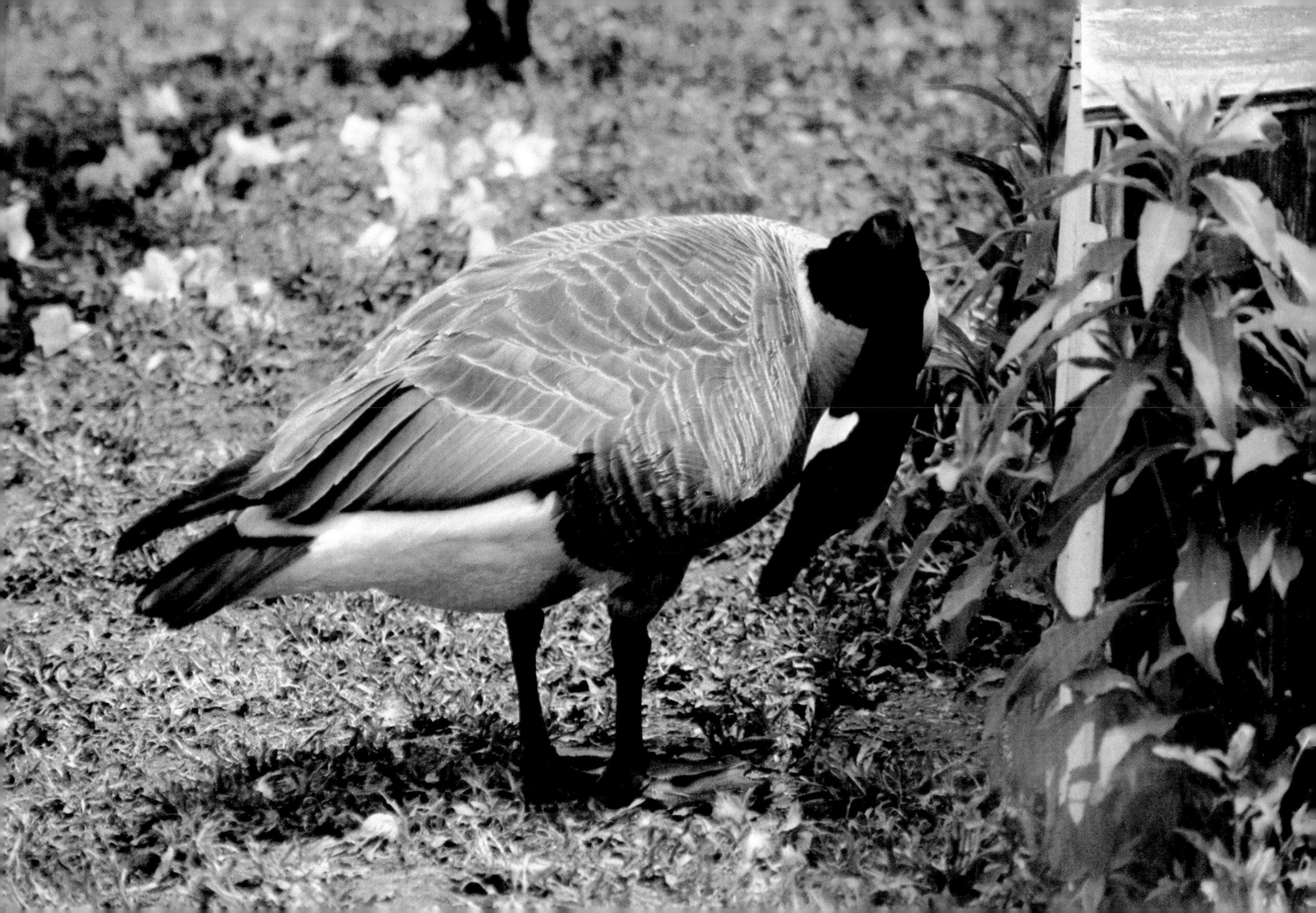

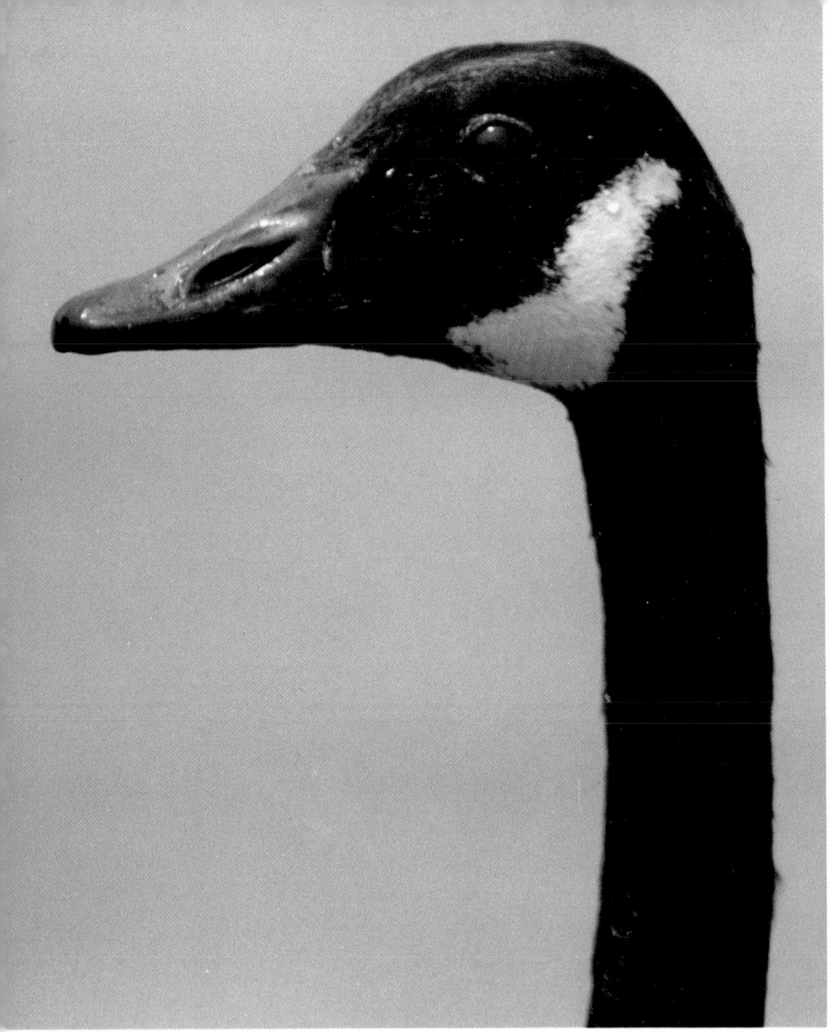

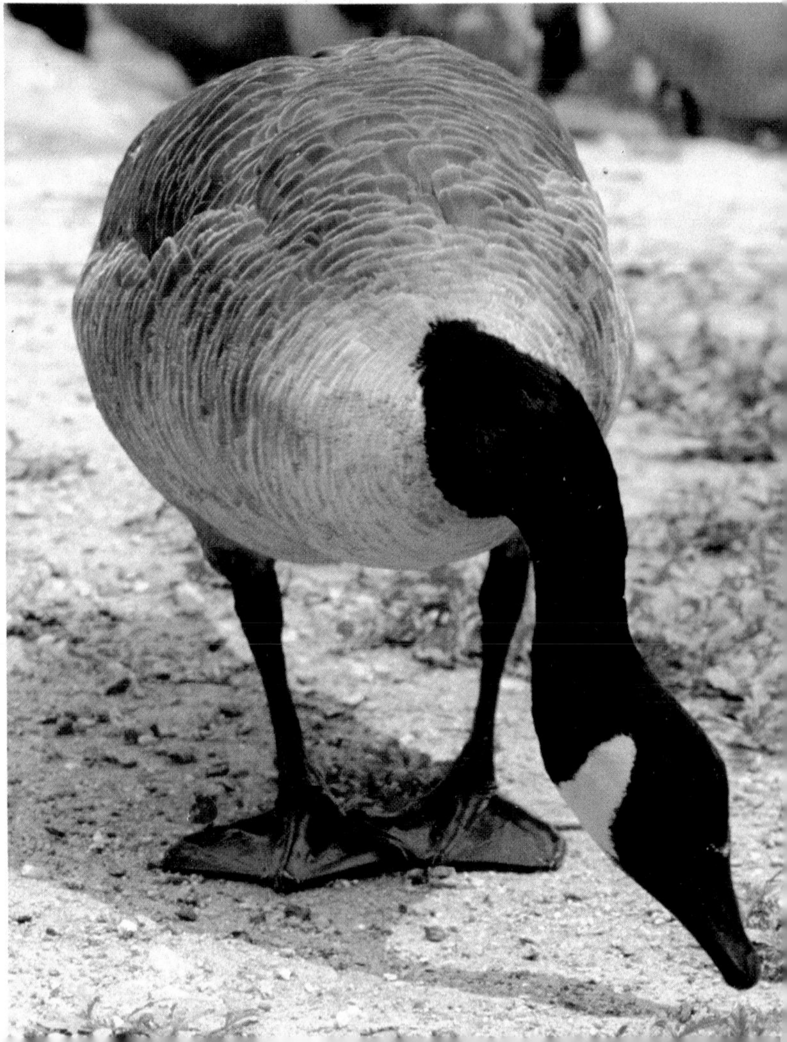

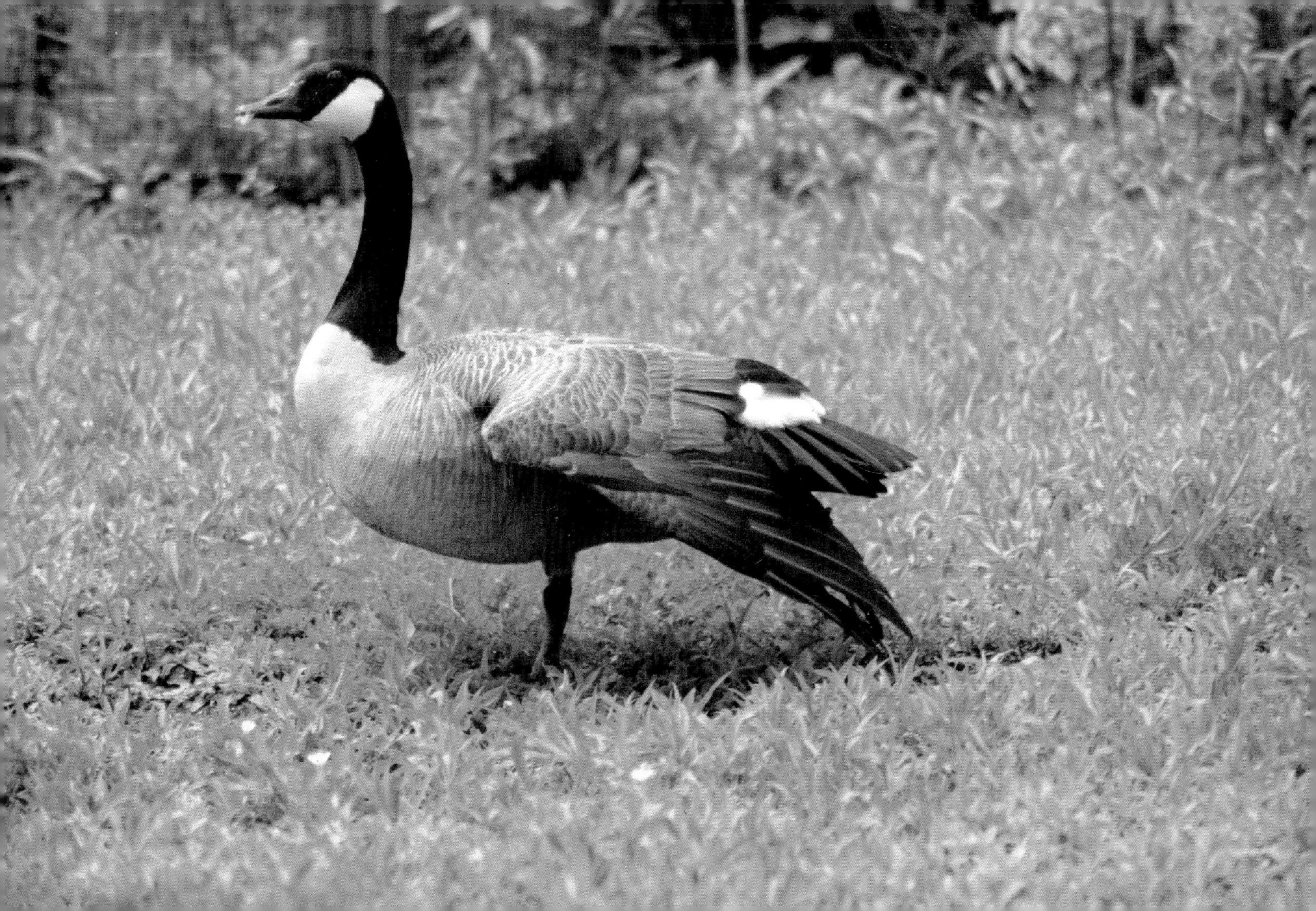

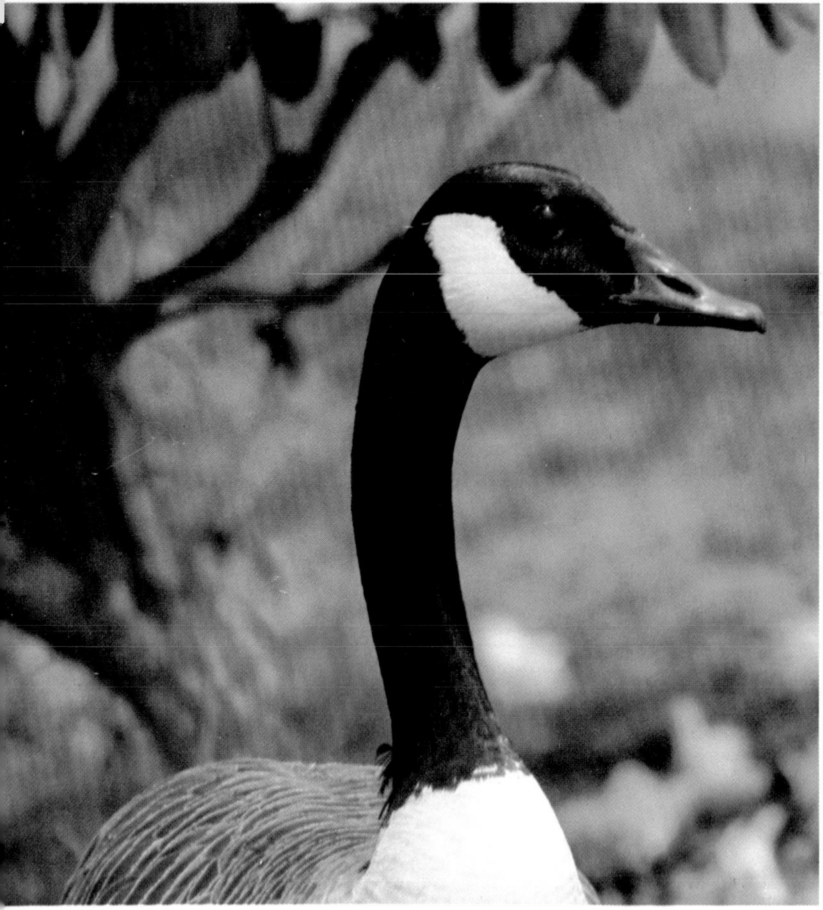
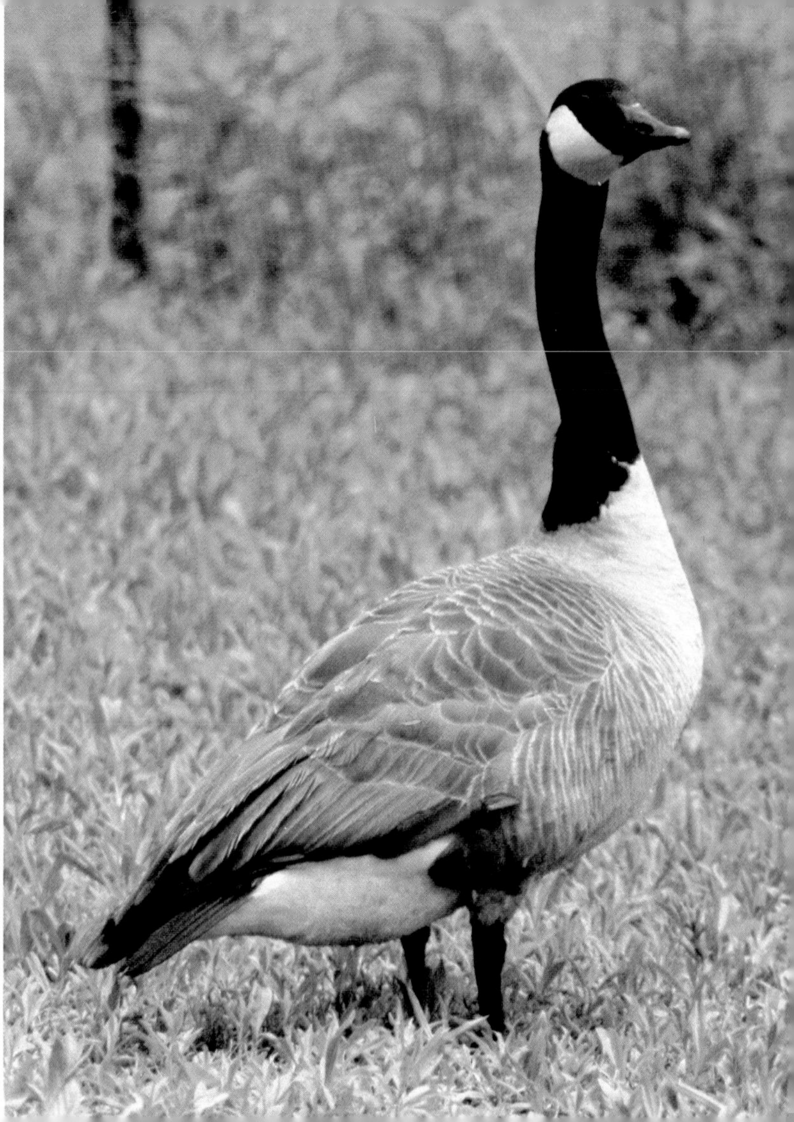

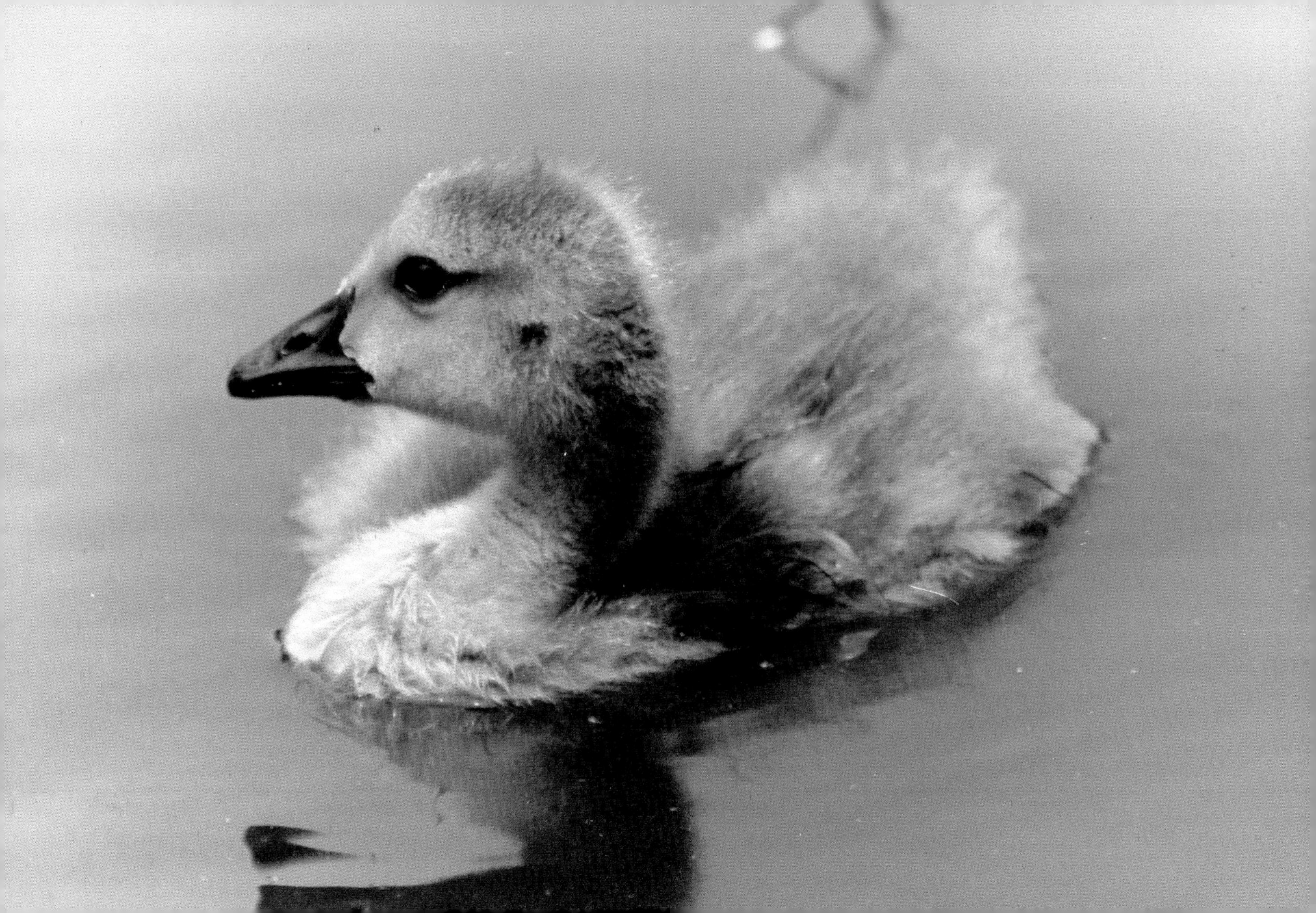

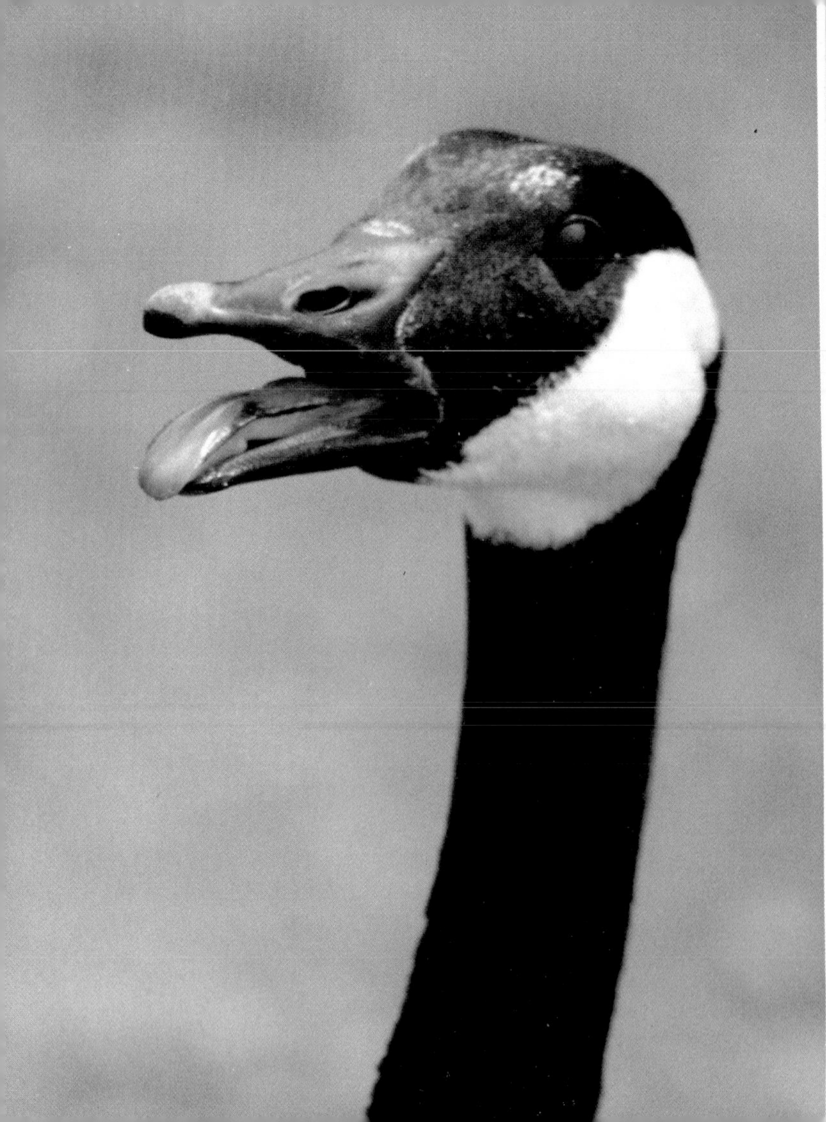

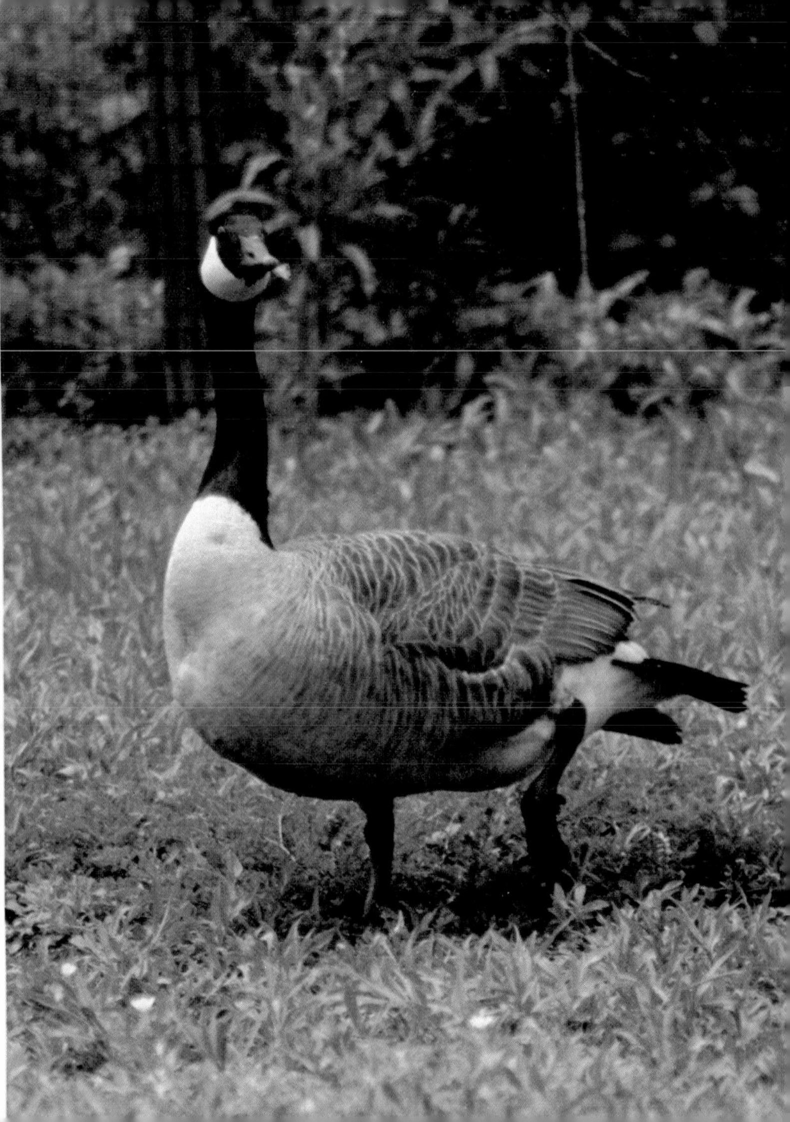

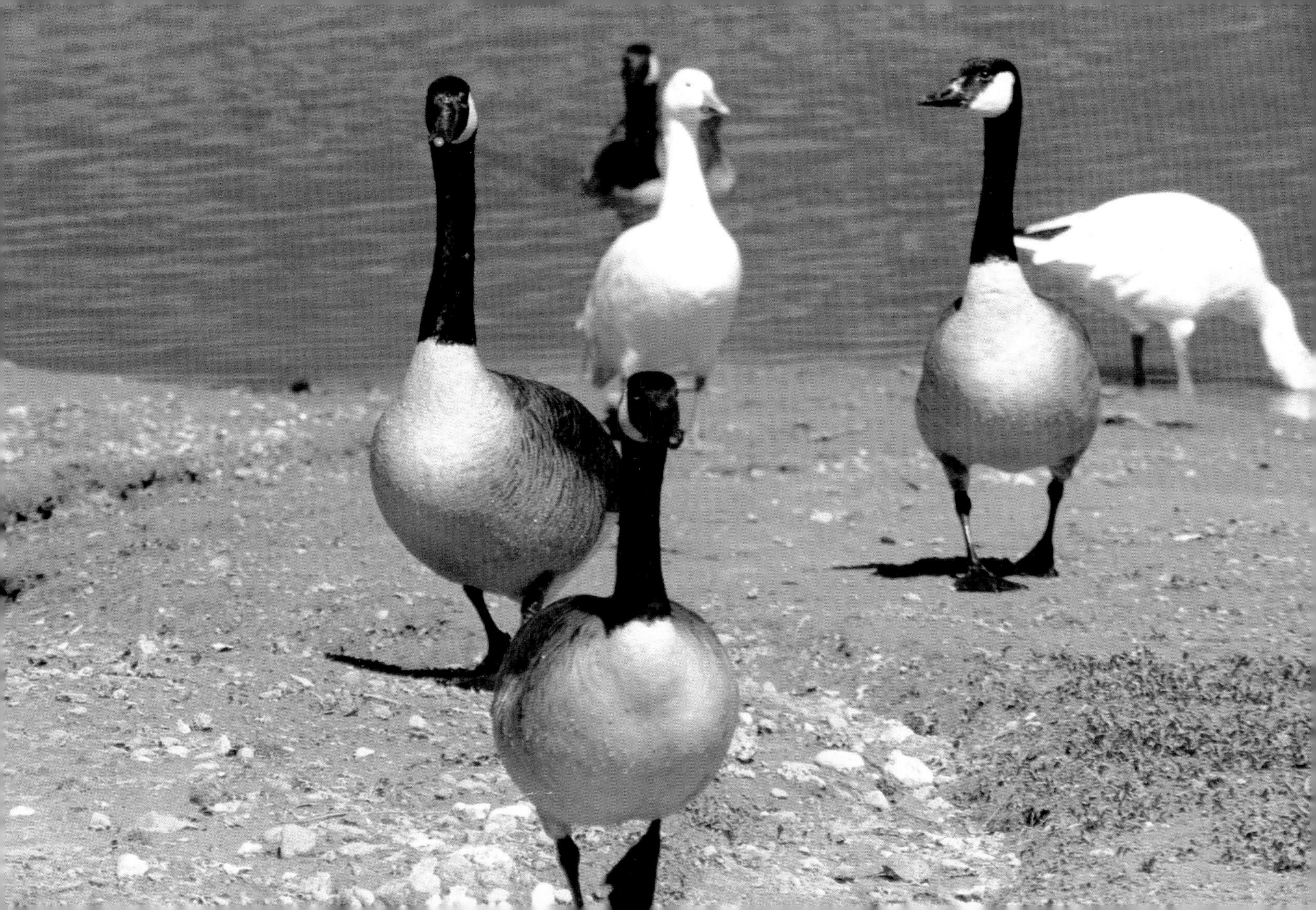

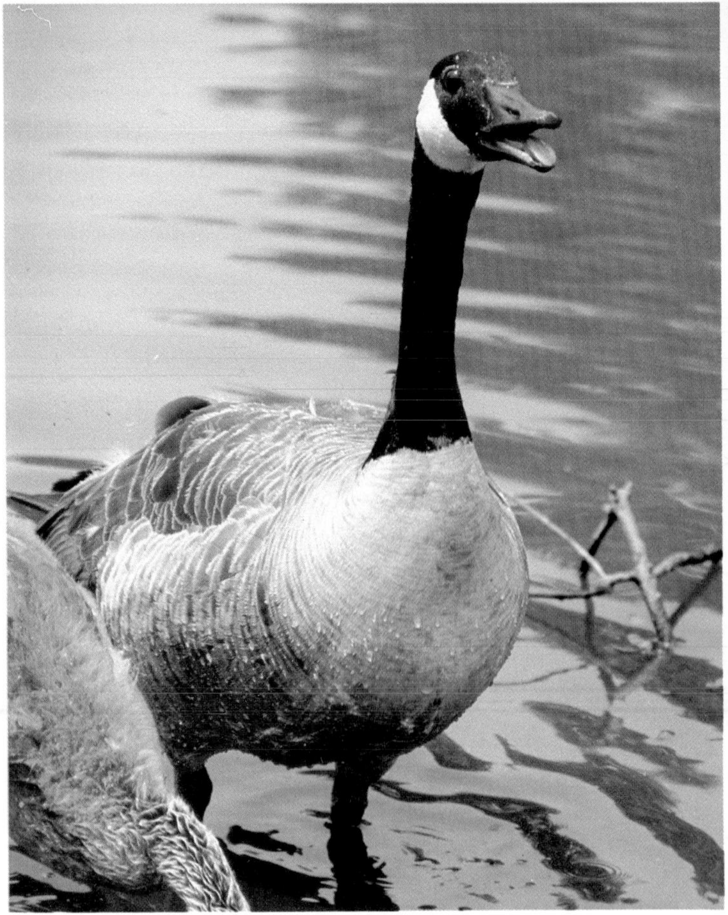

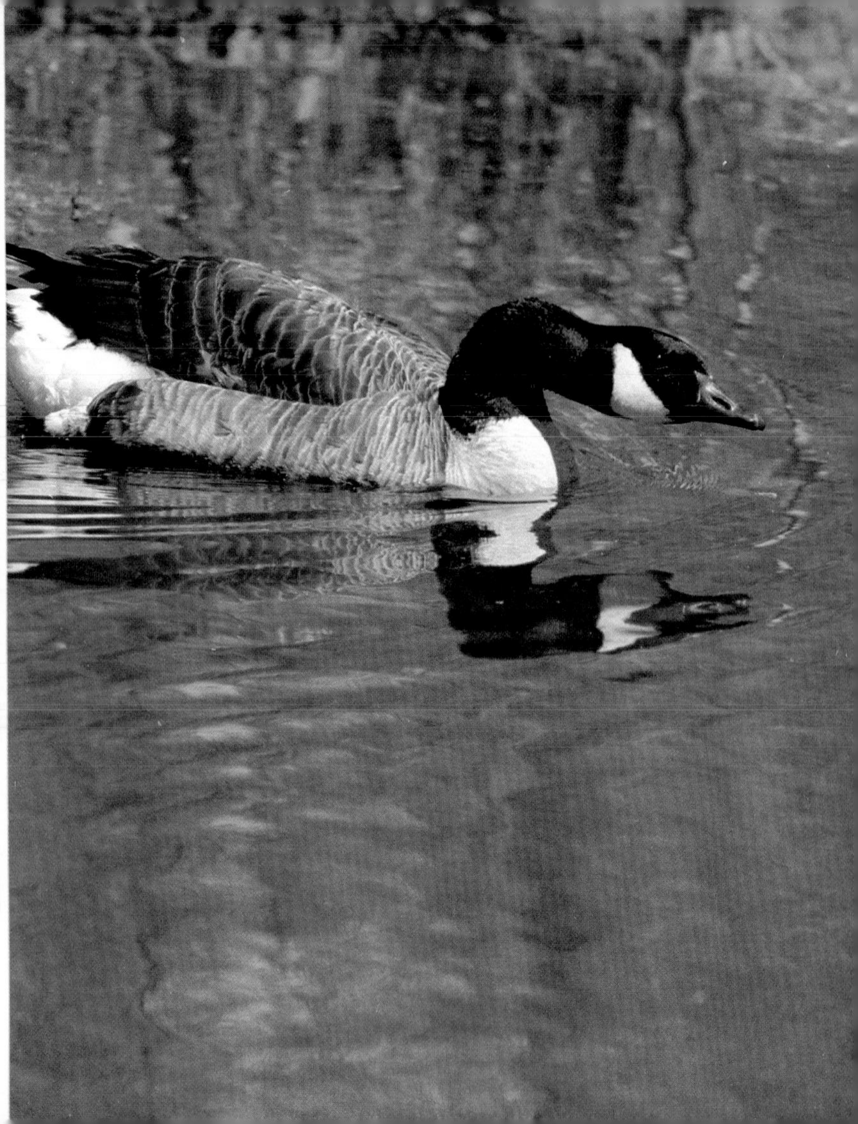

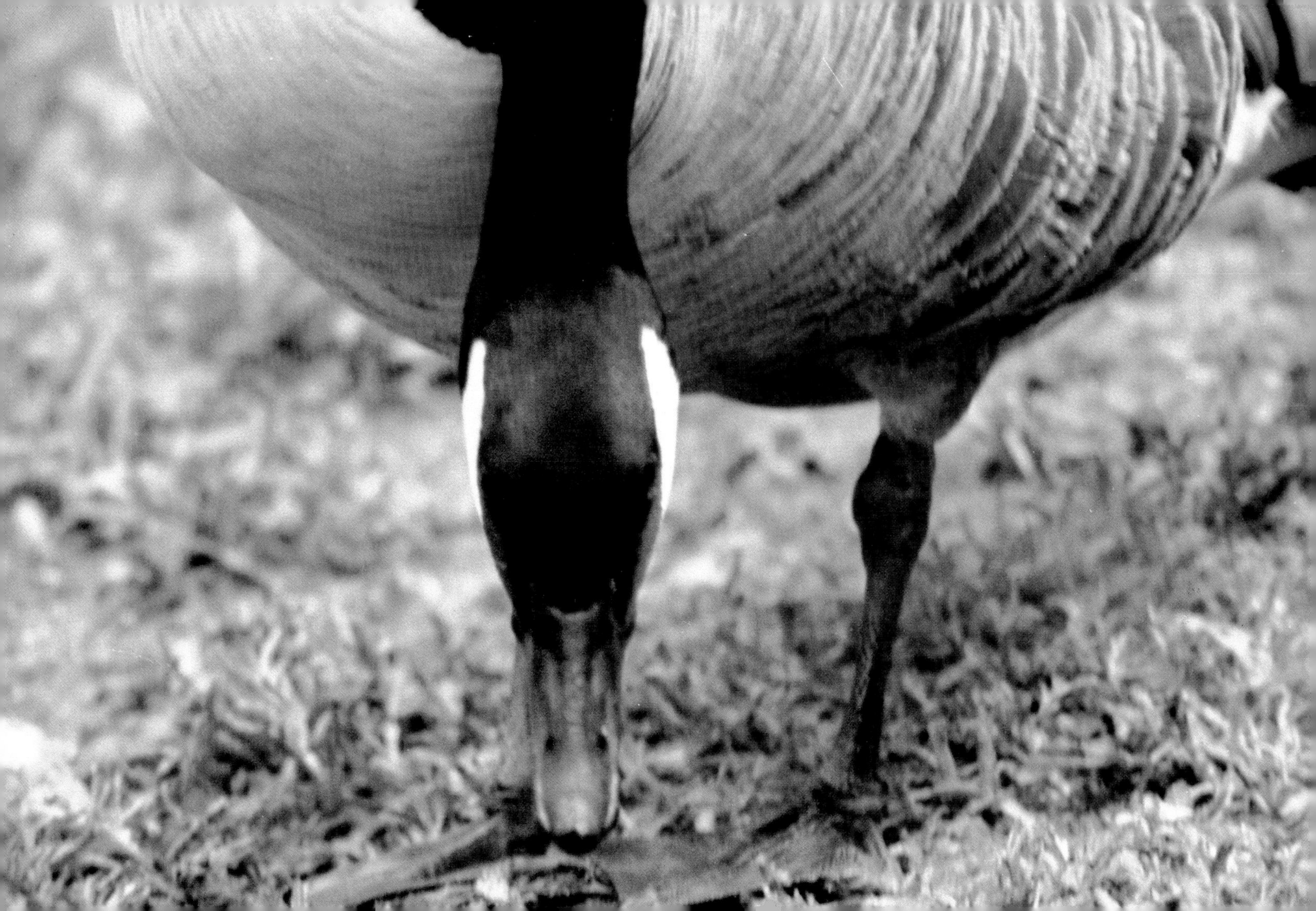

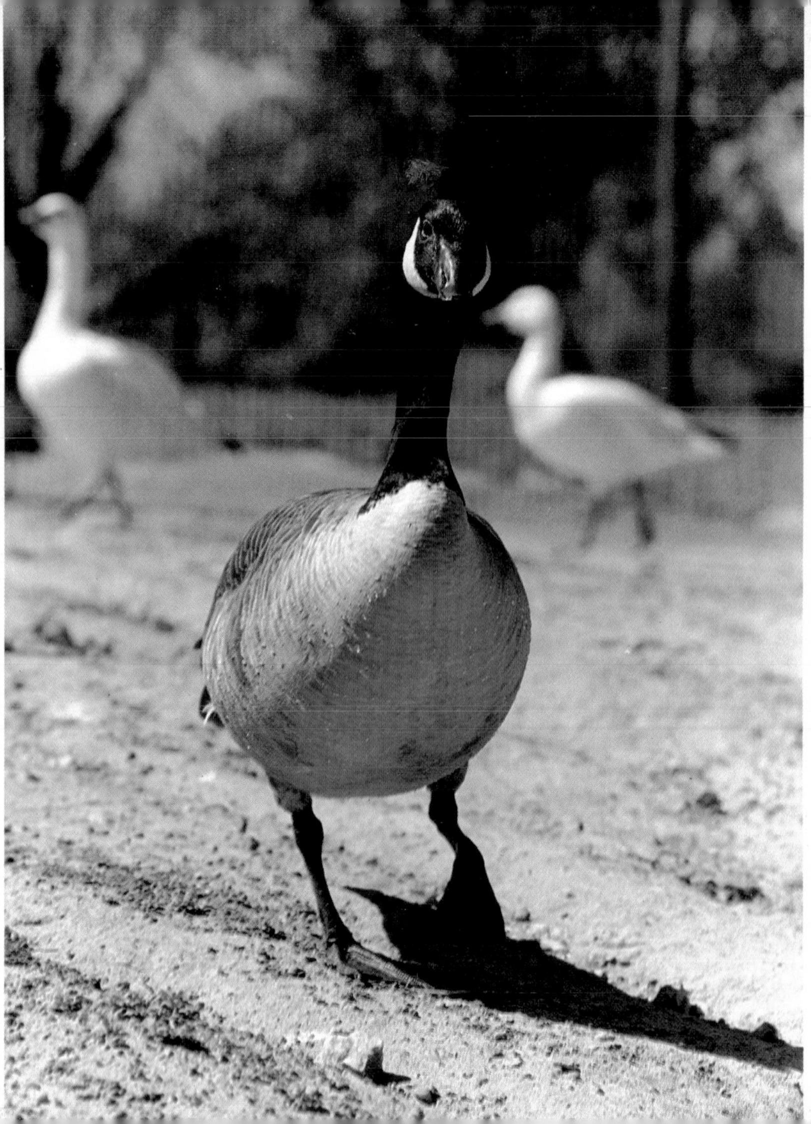

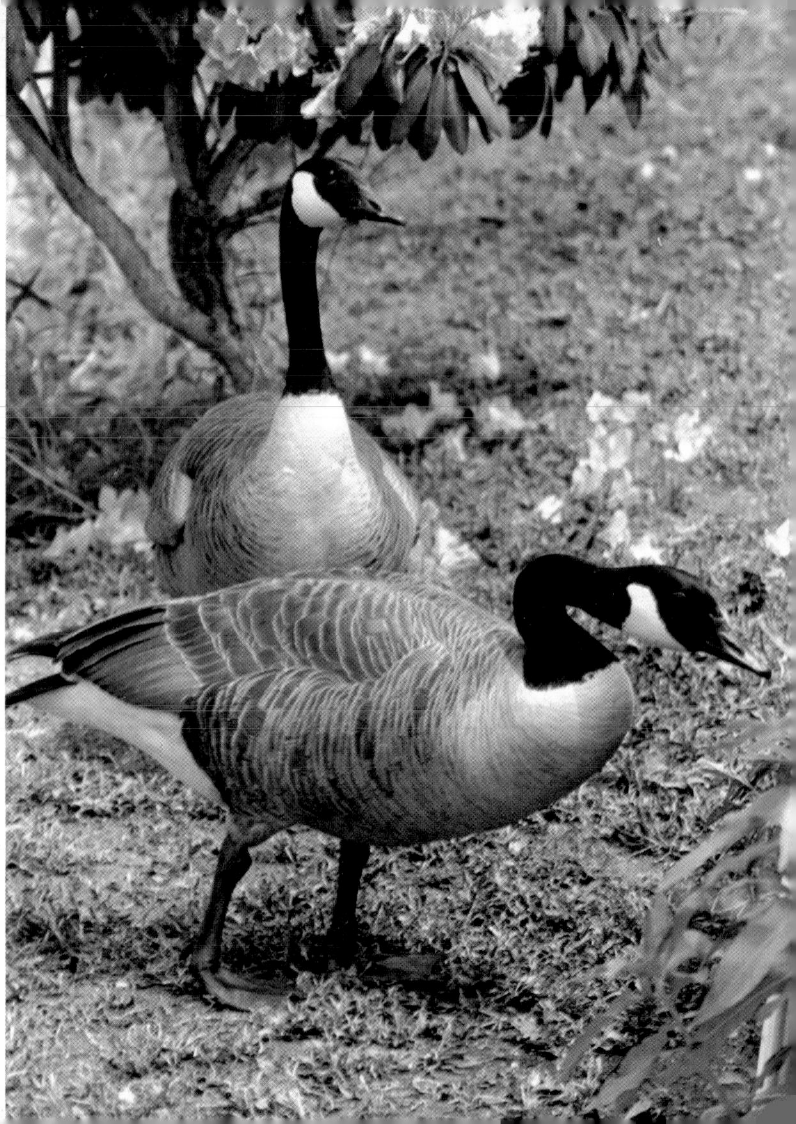

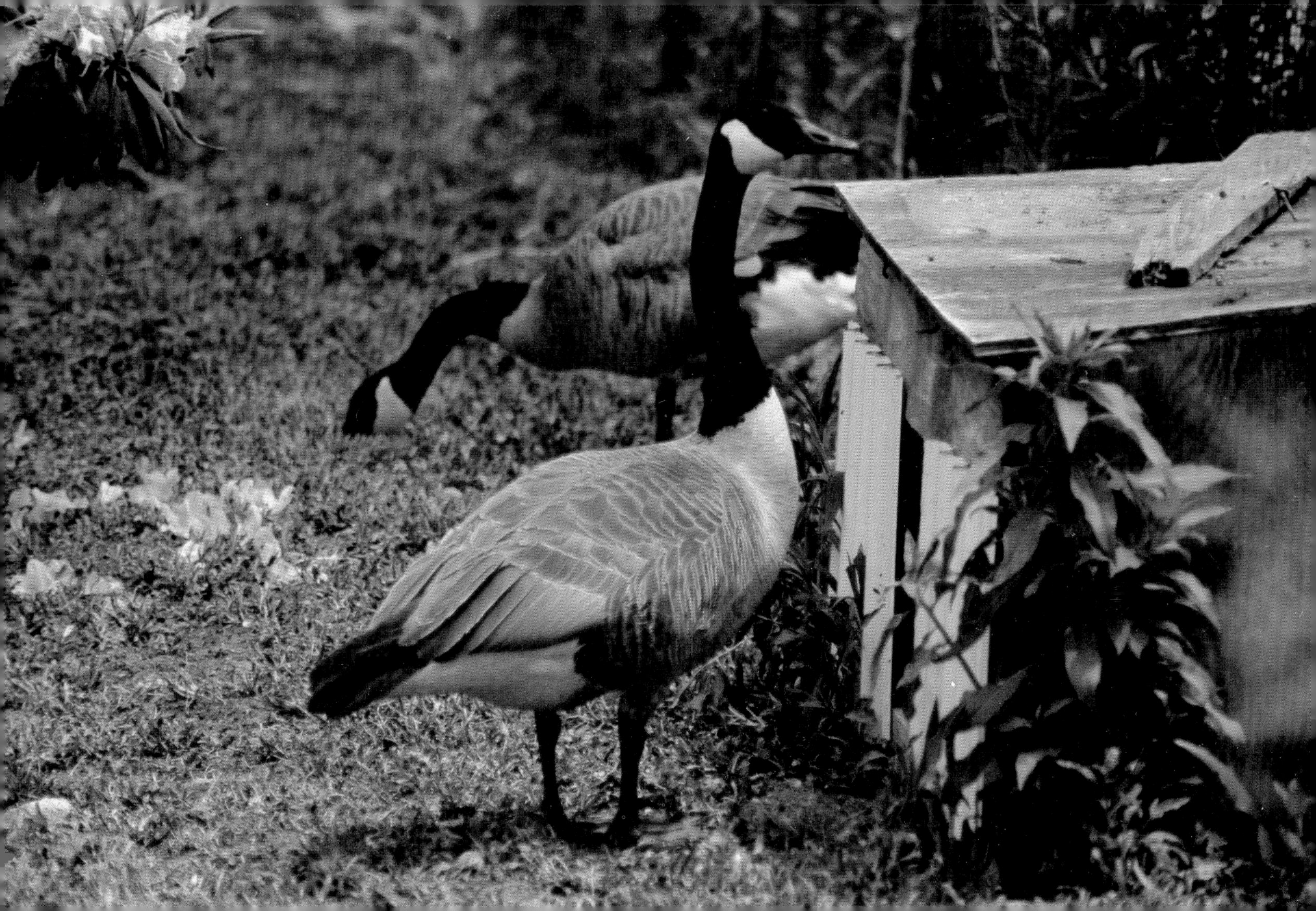

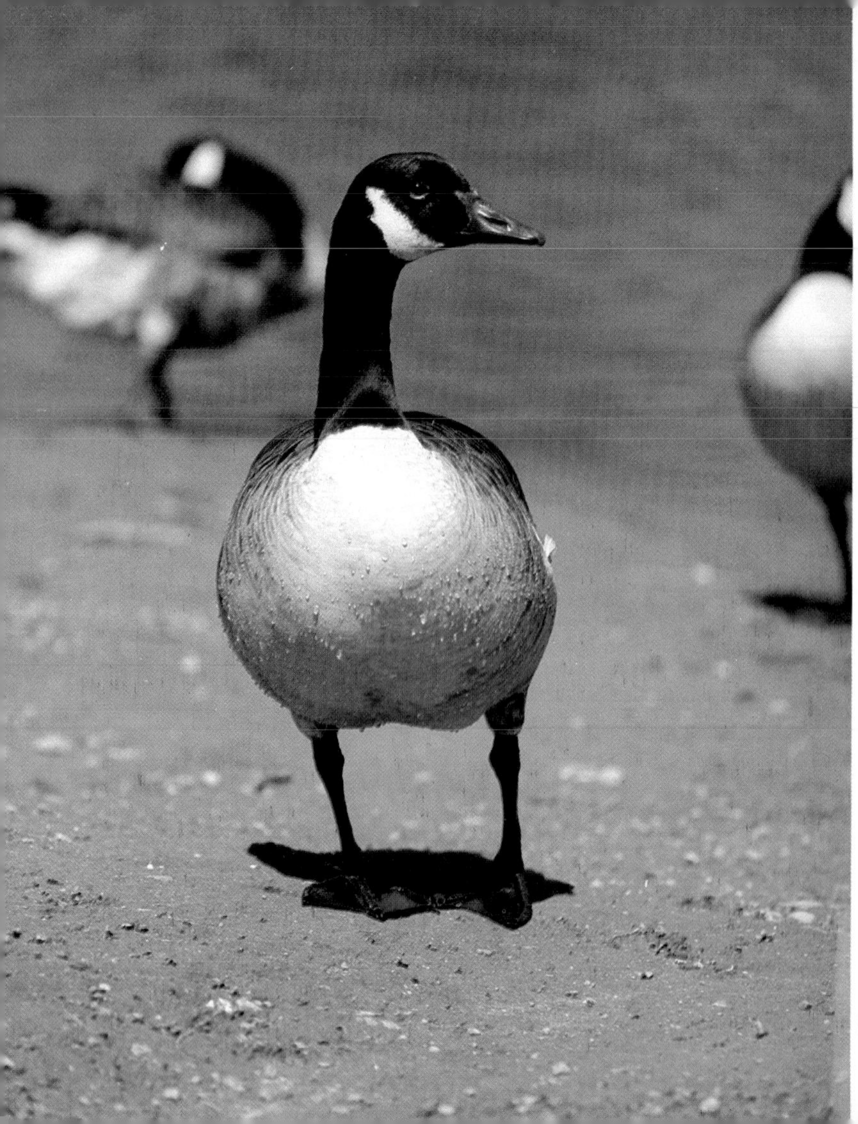
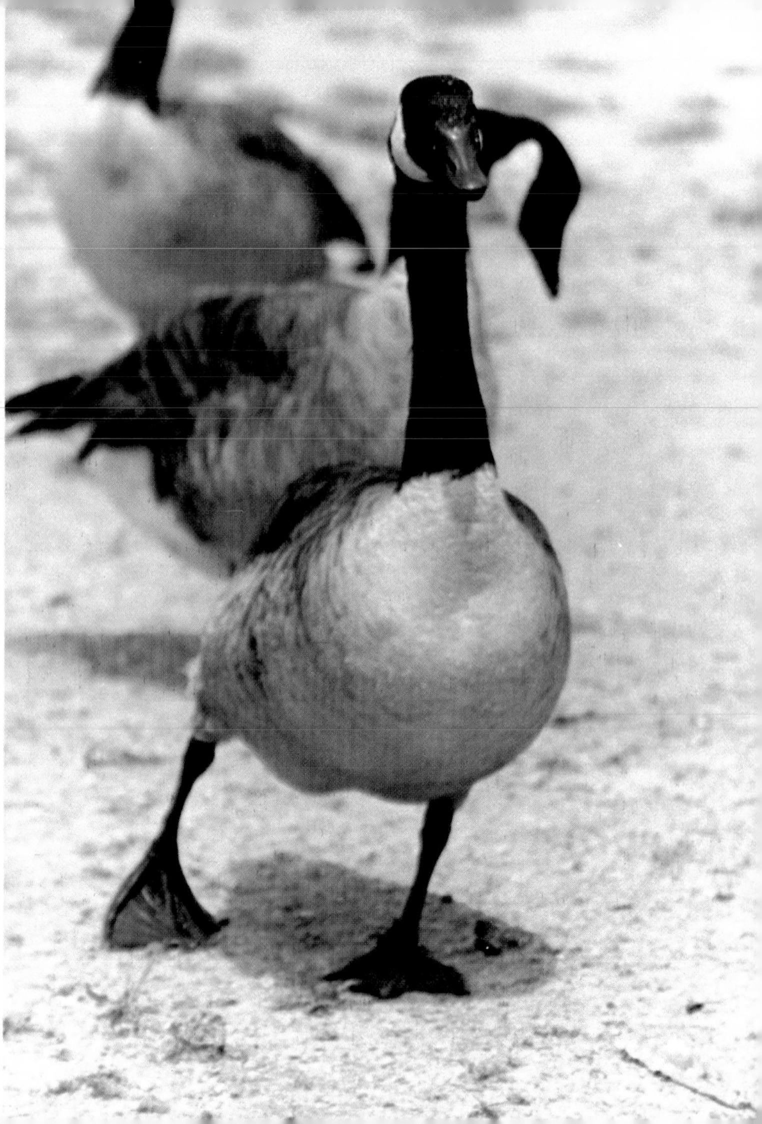

Richards' & Cackling Geese

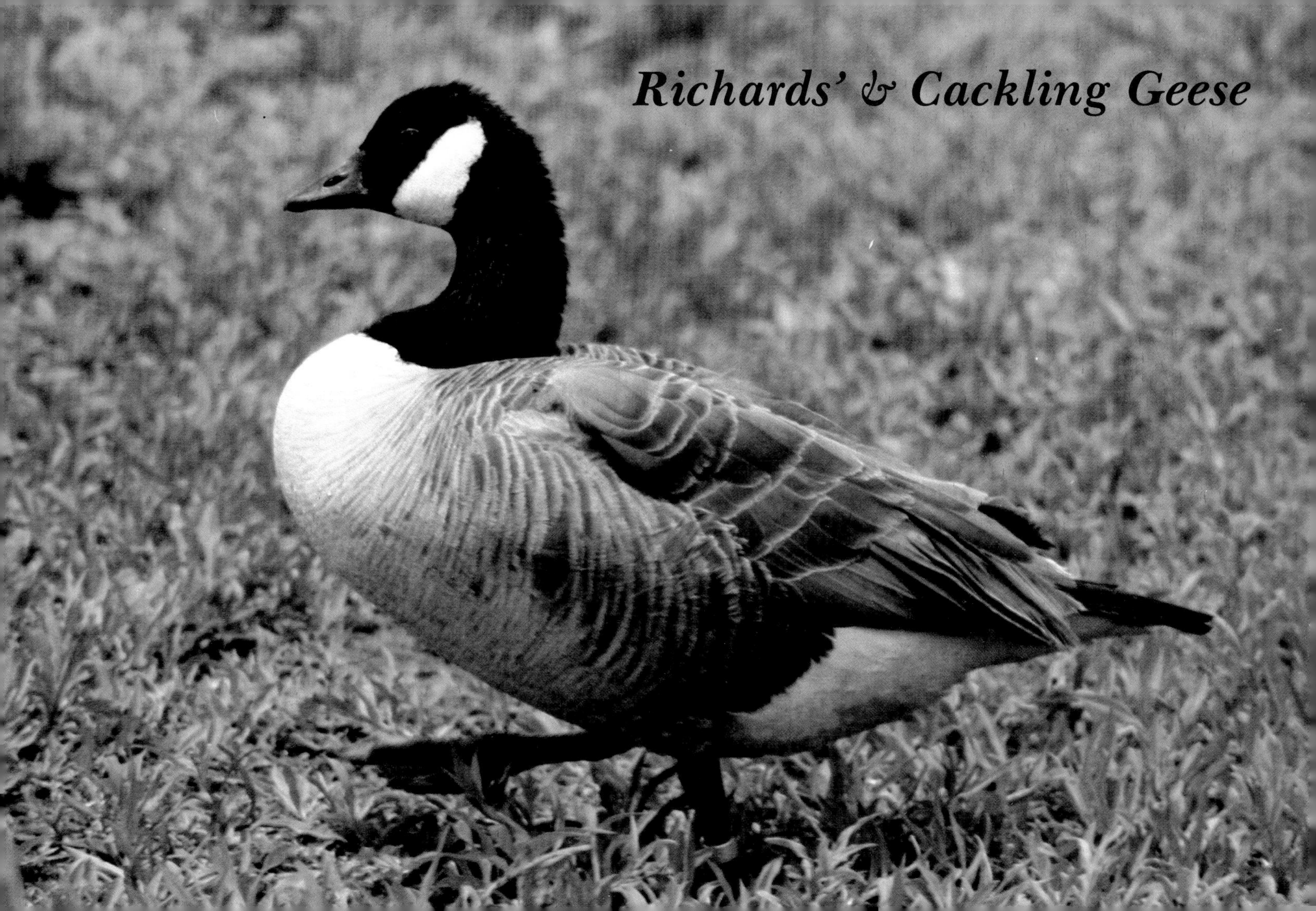

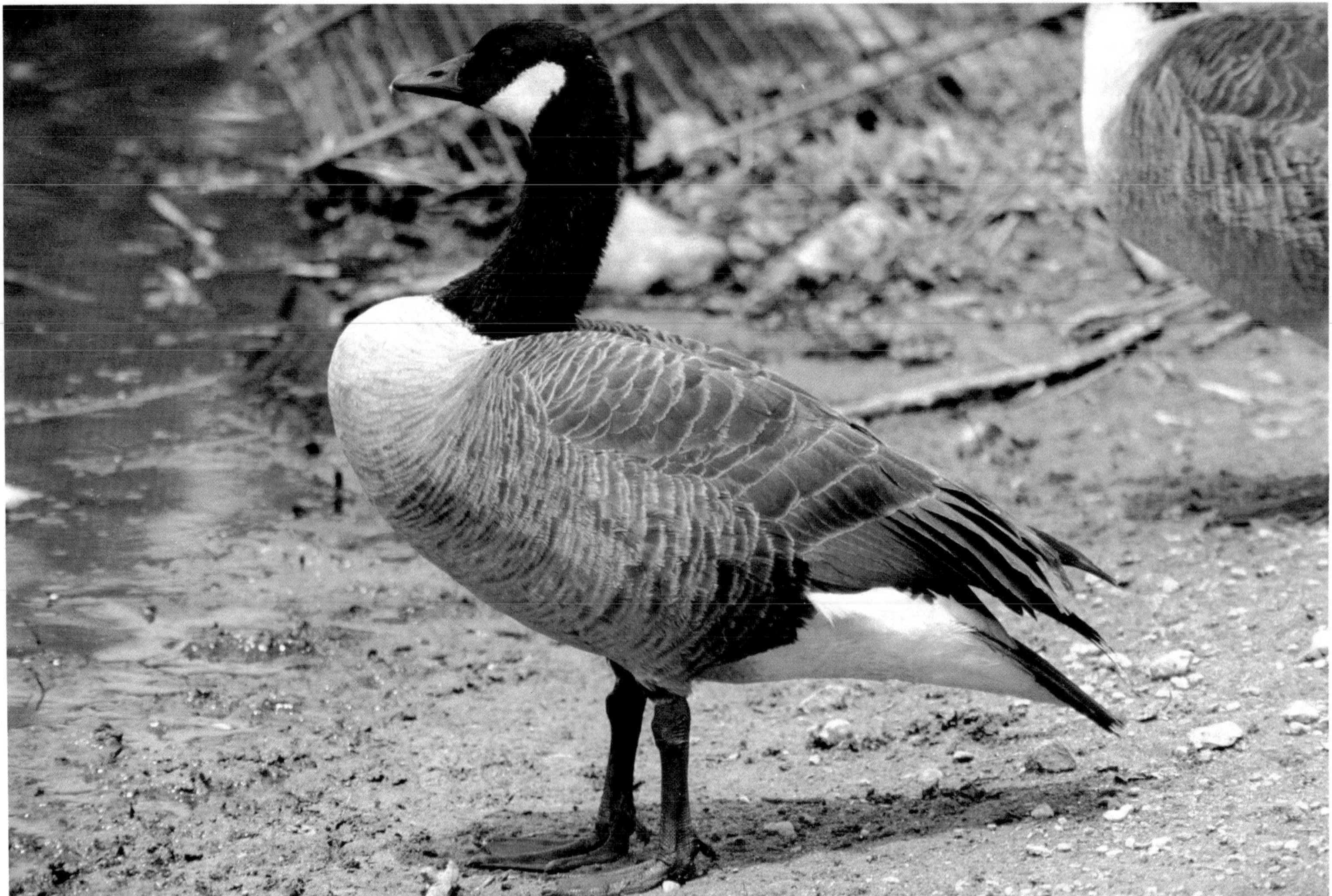

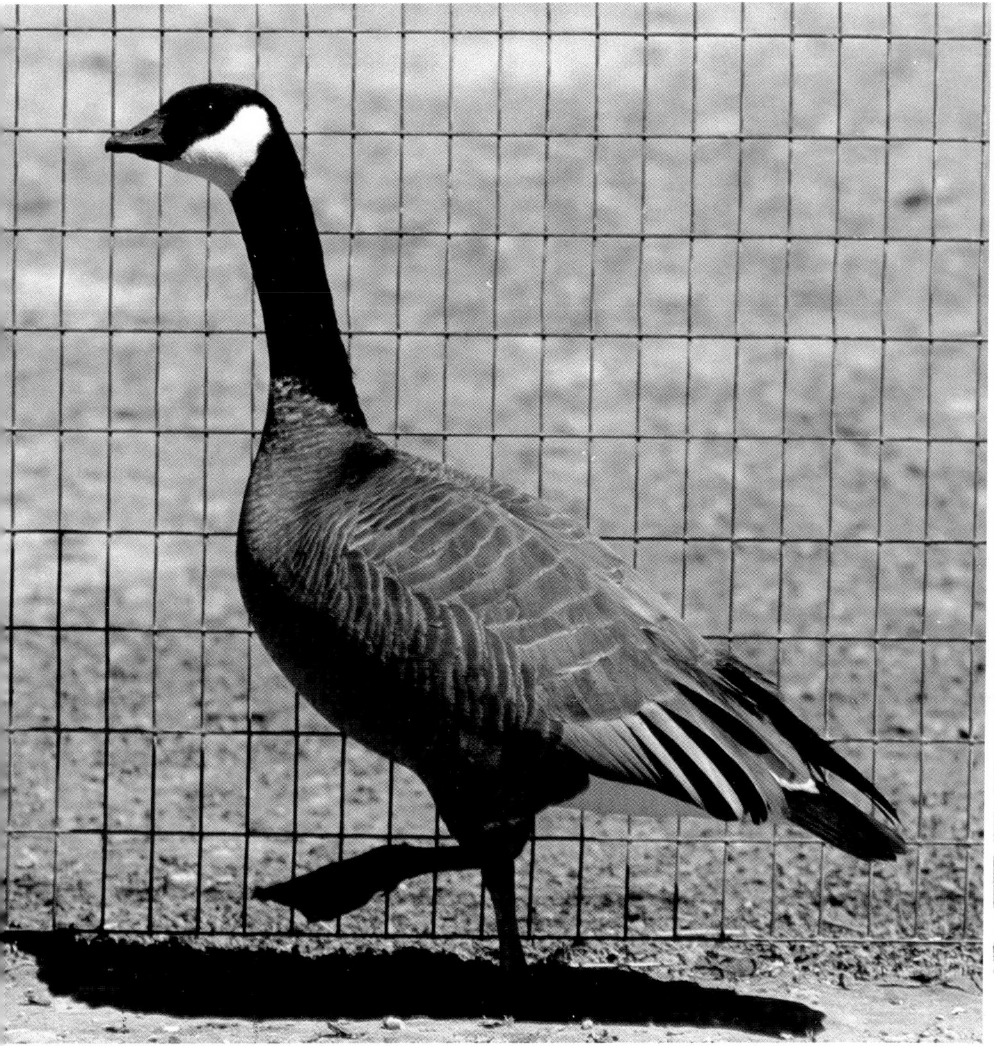
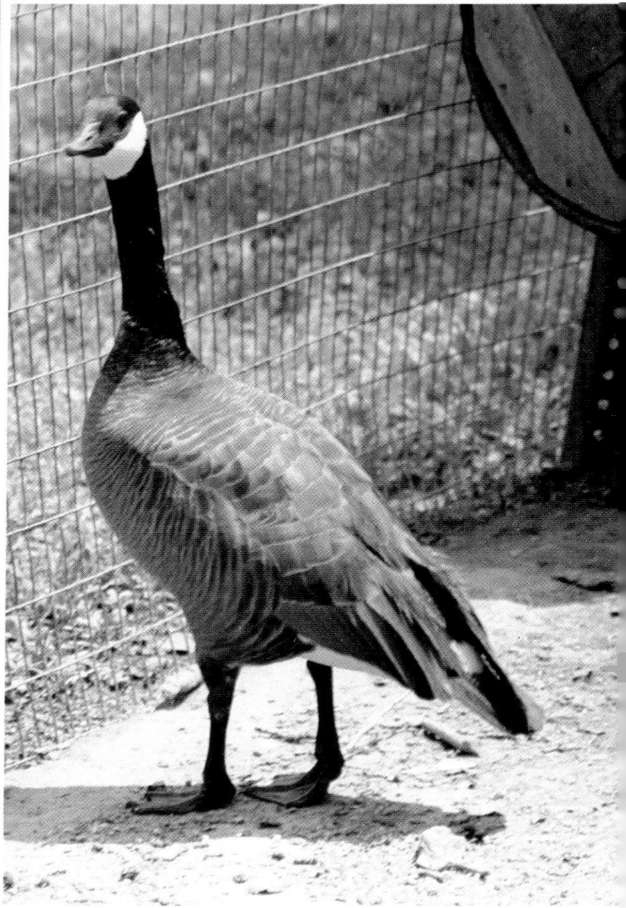

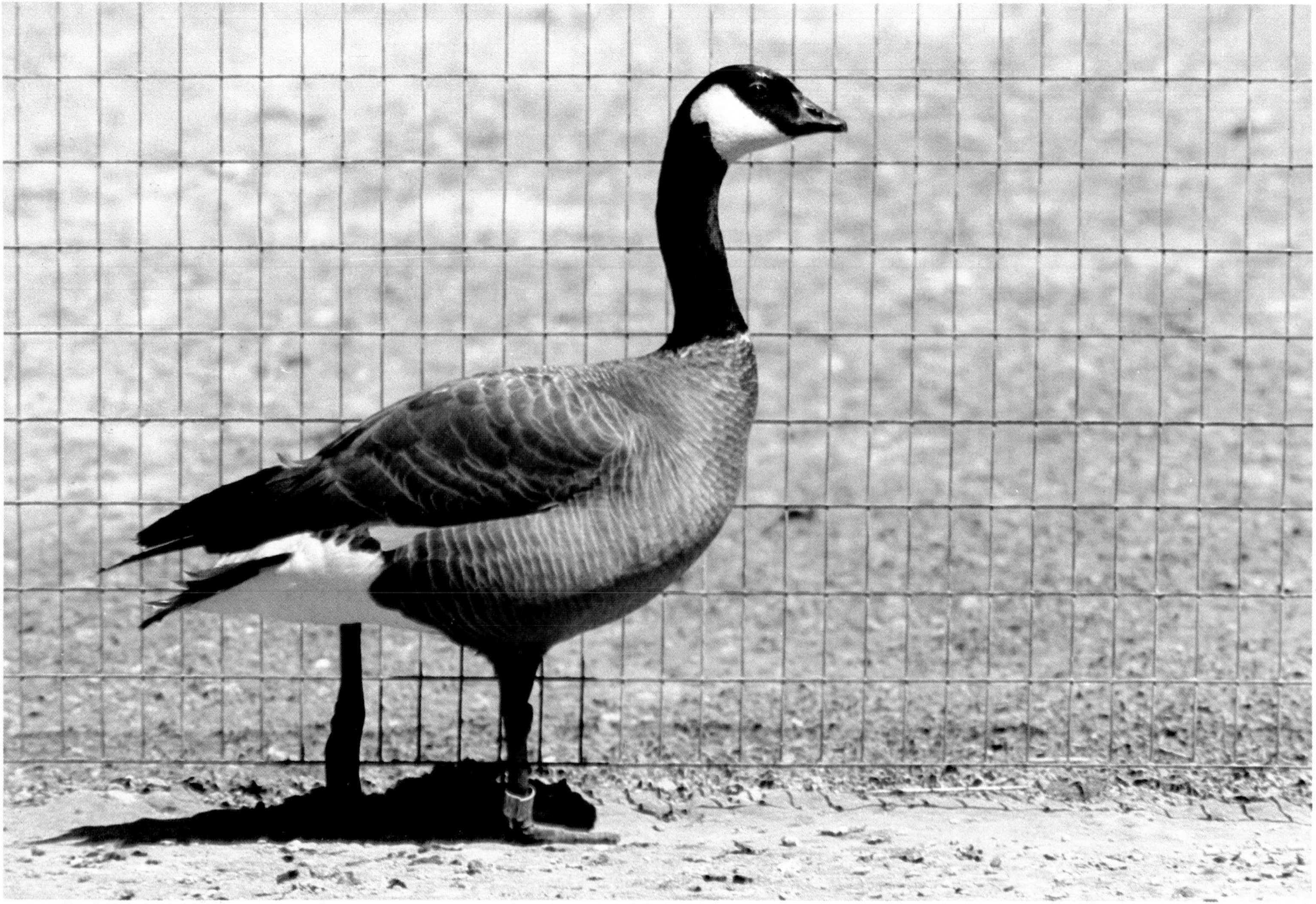

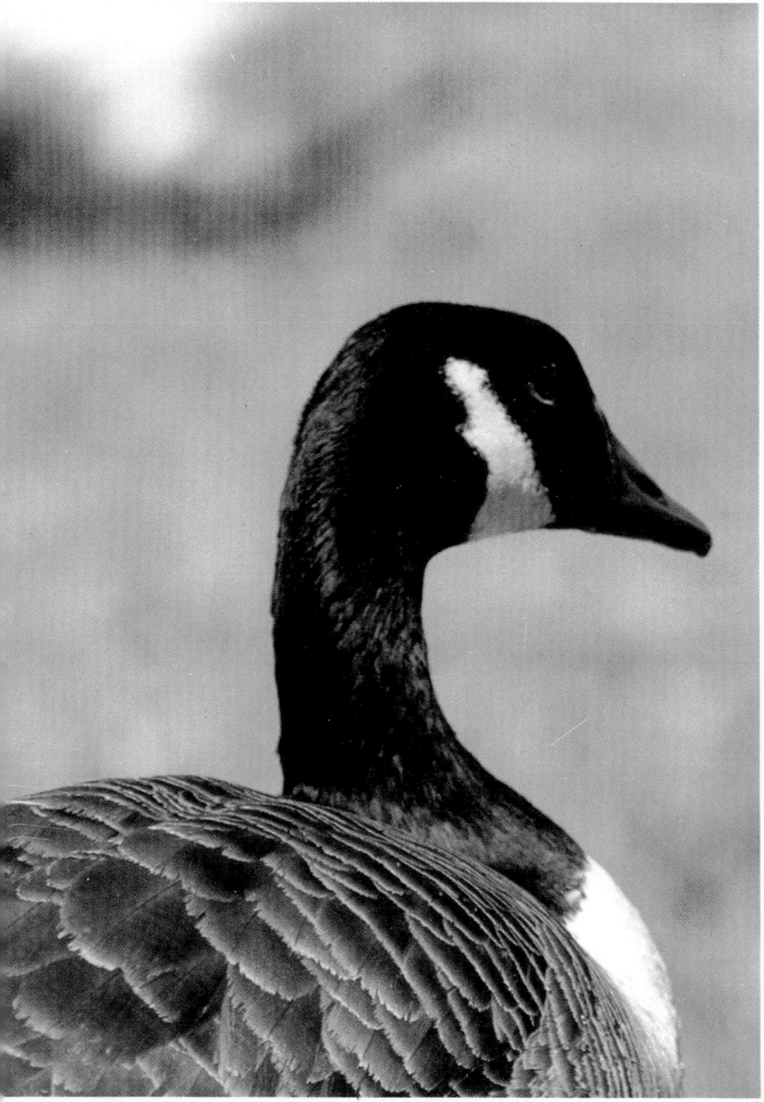

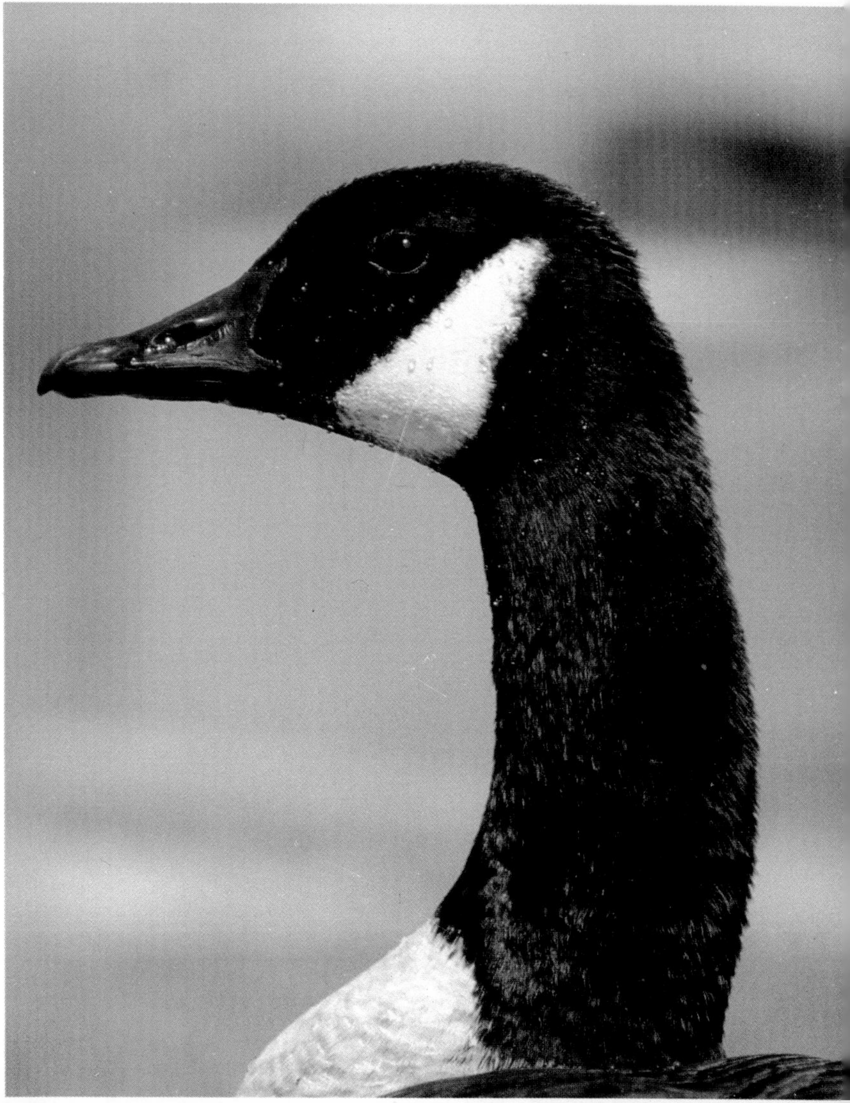

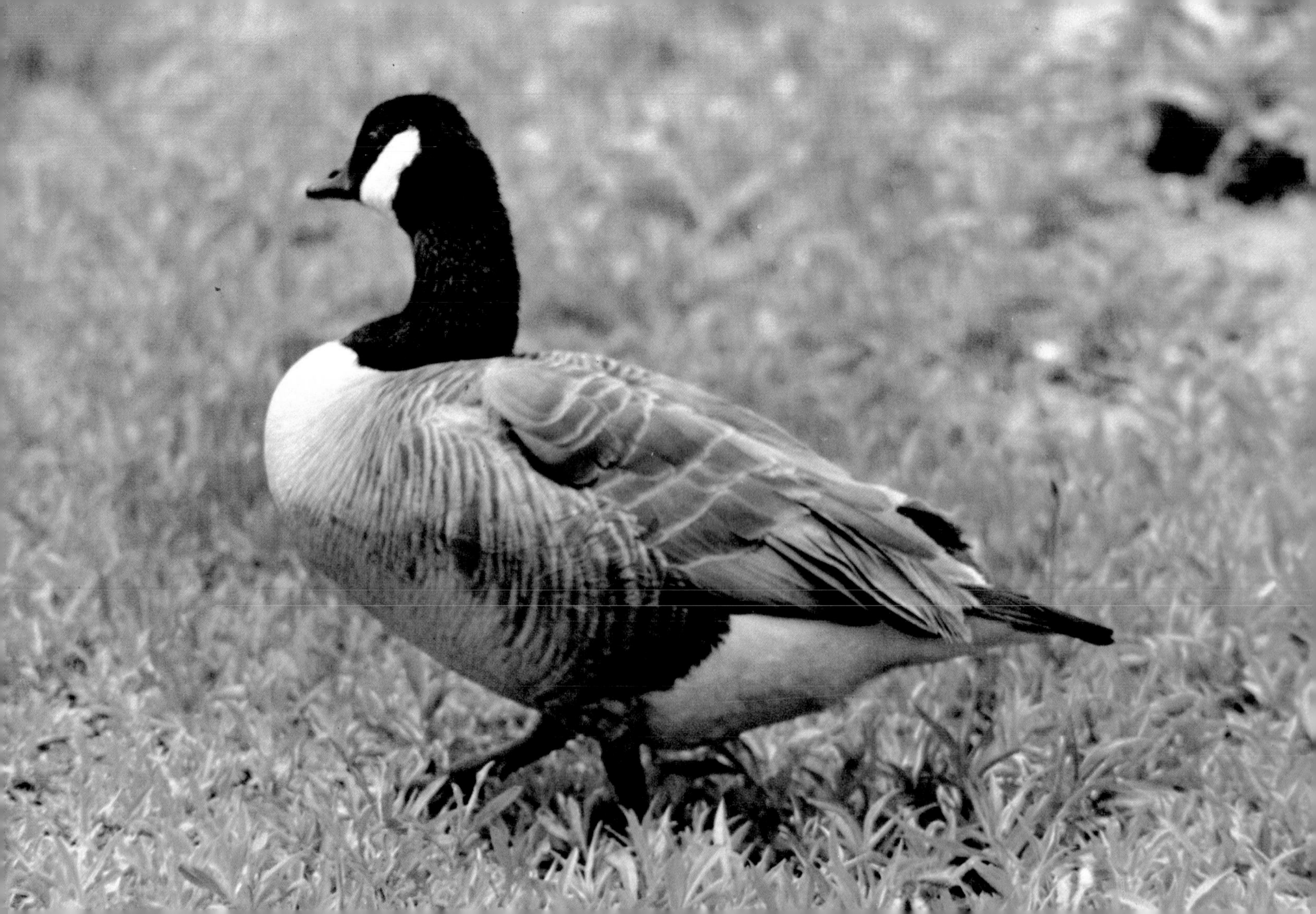

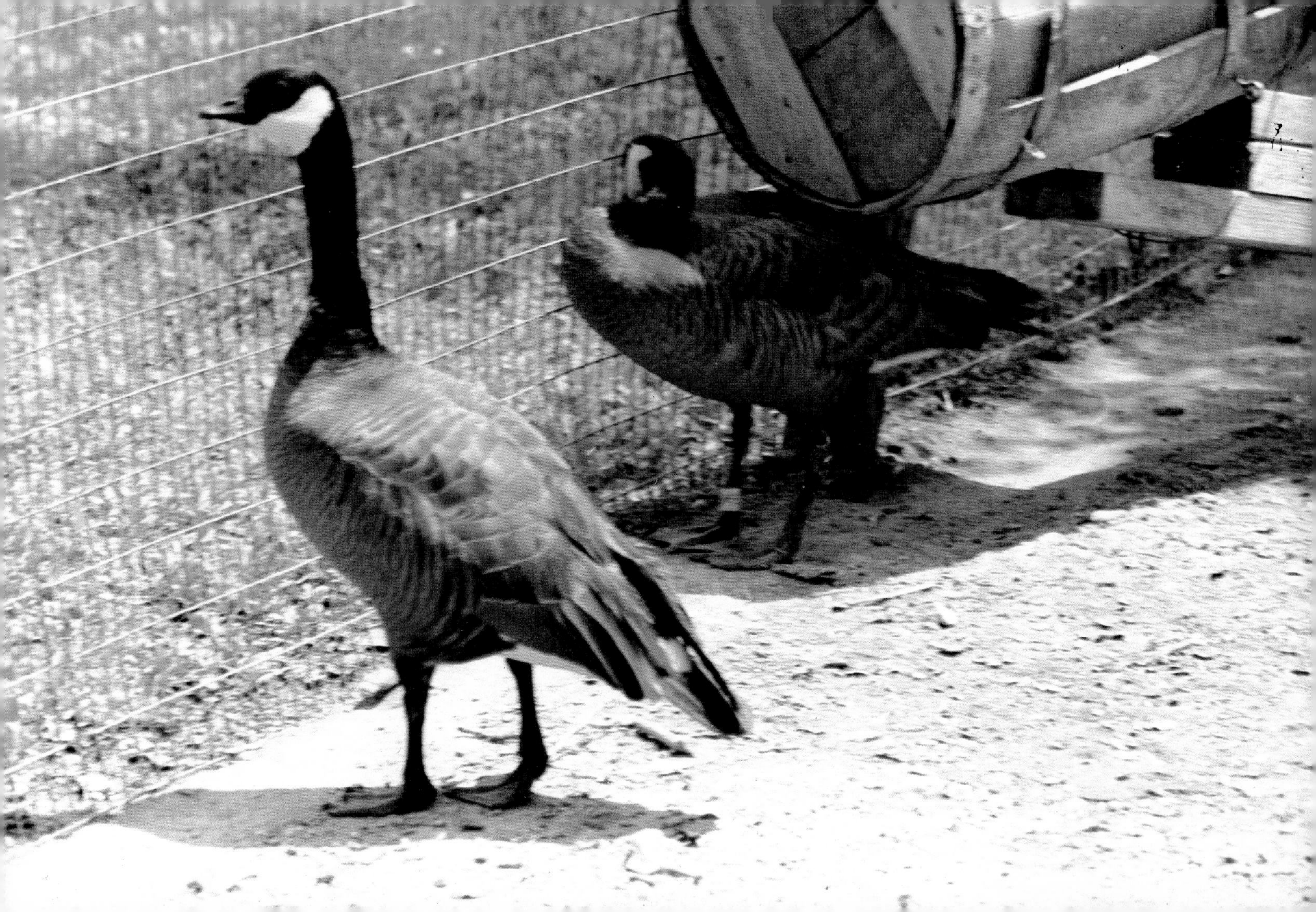

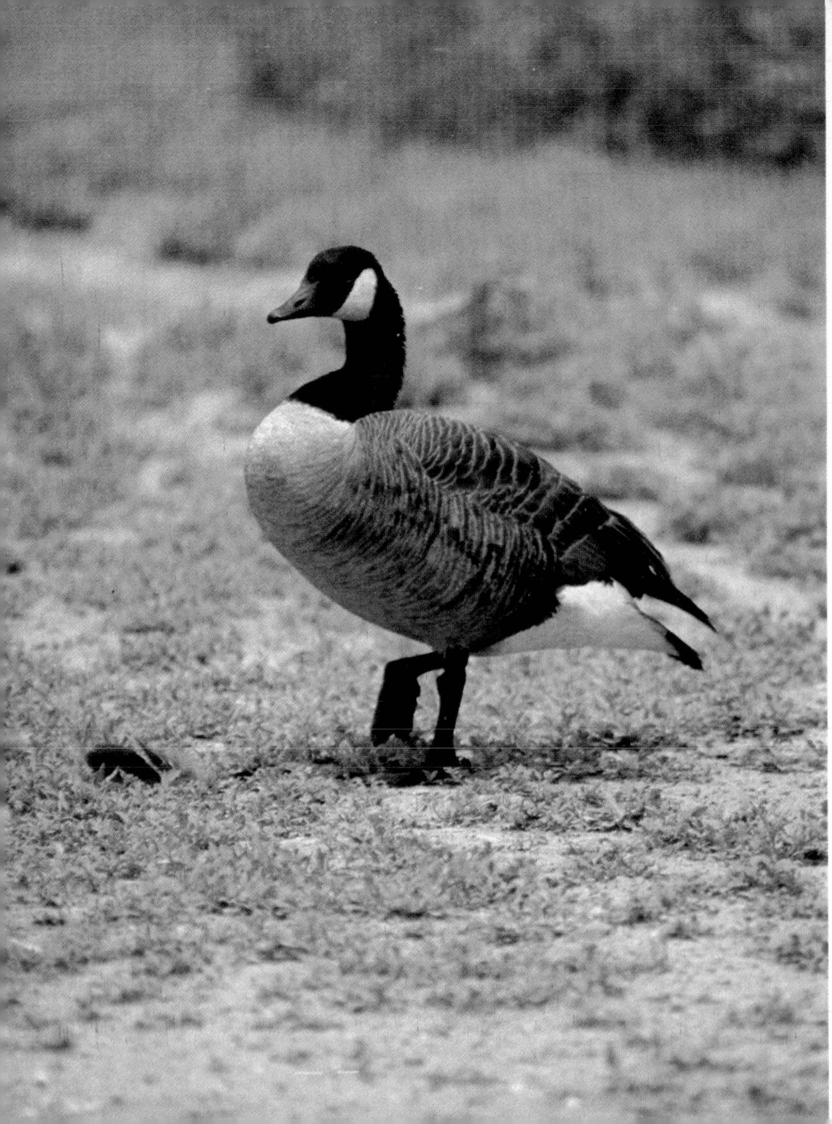
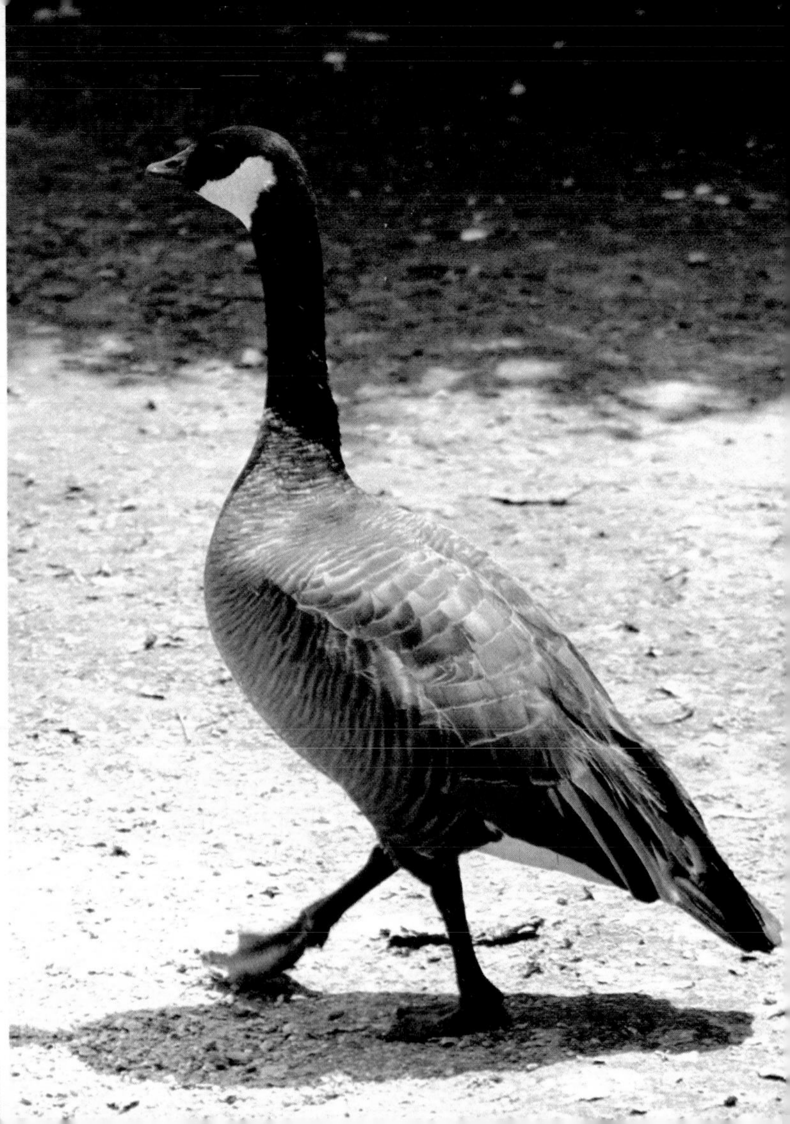

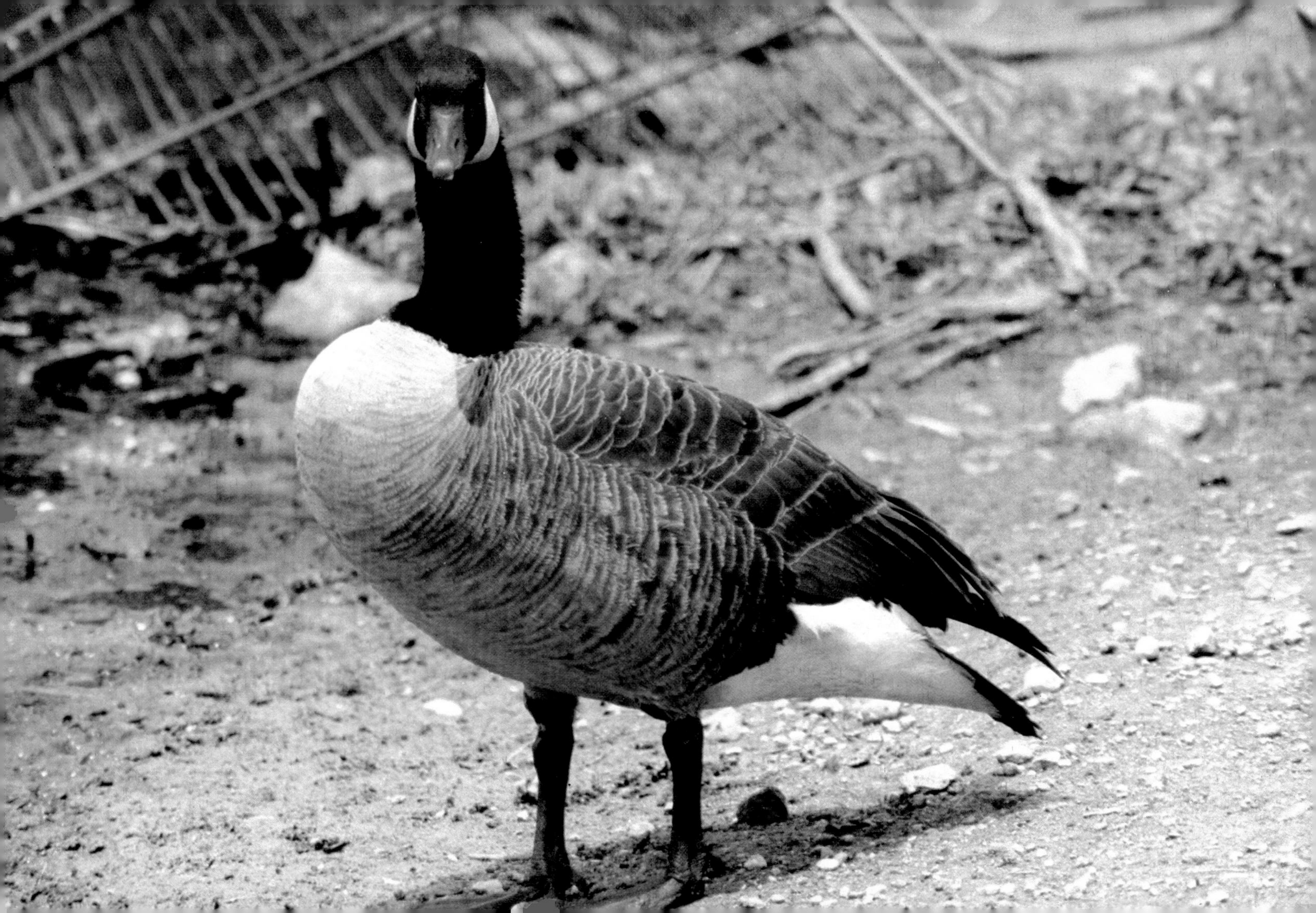

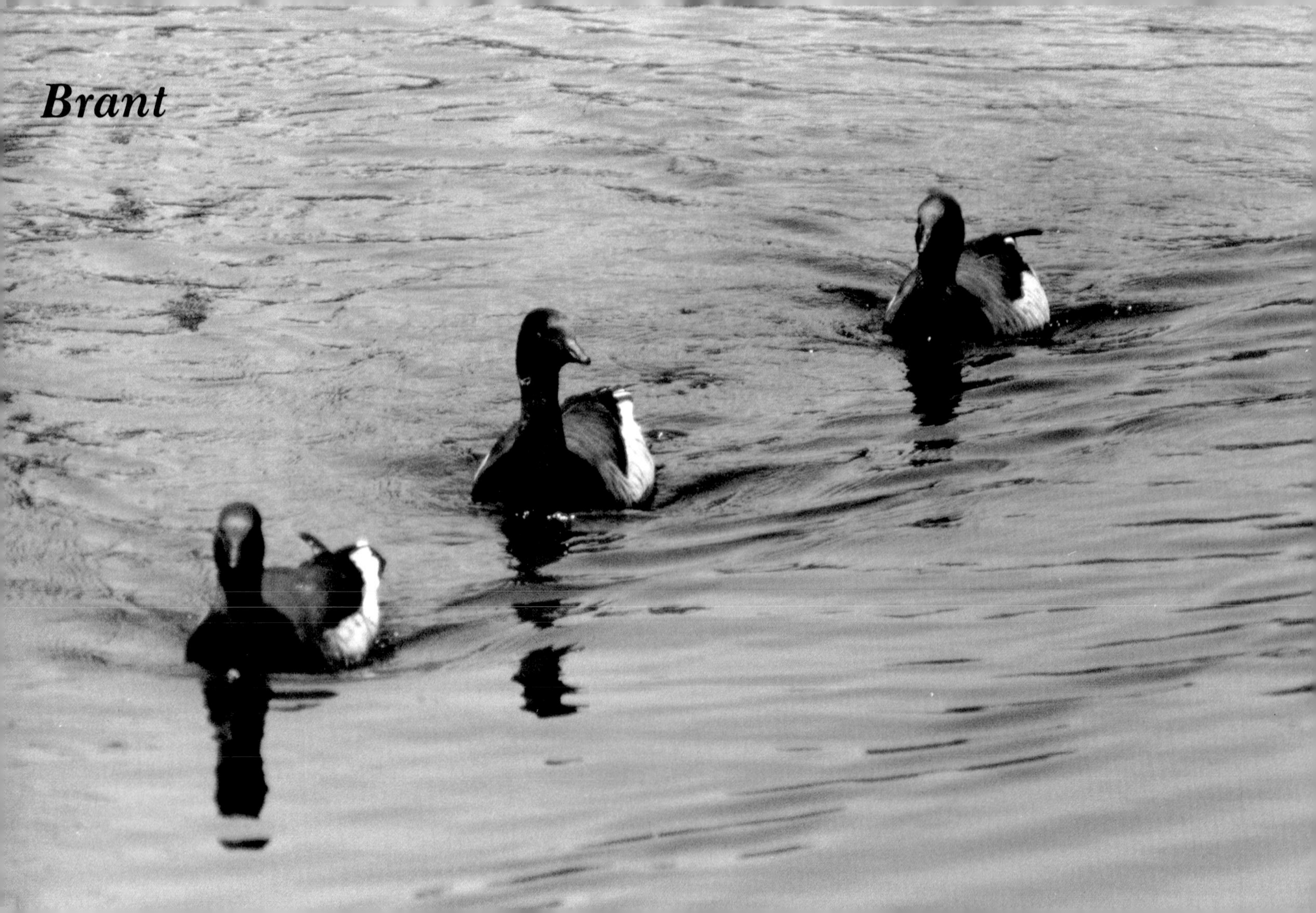
Brant

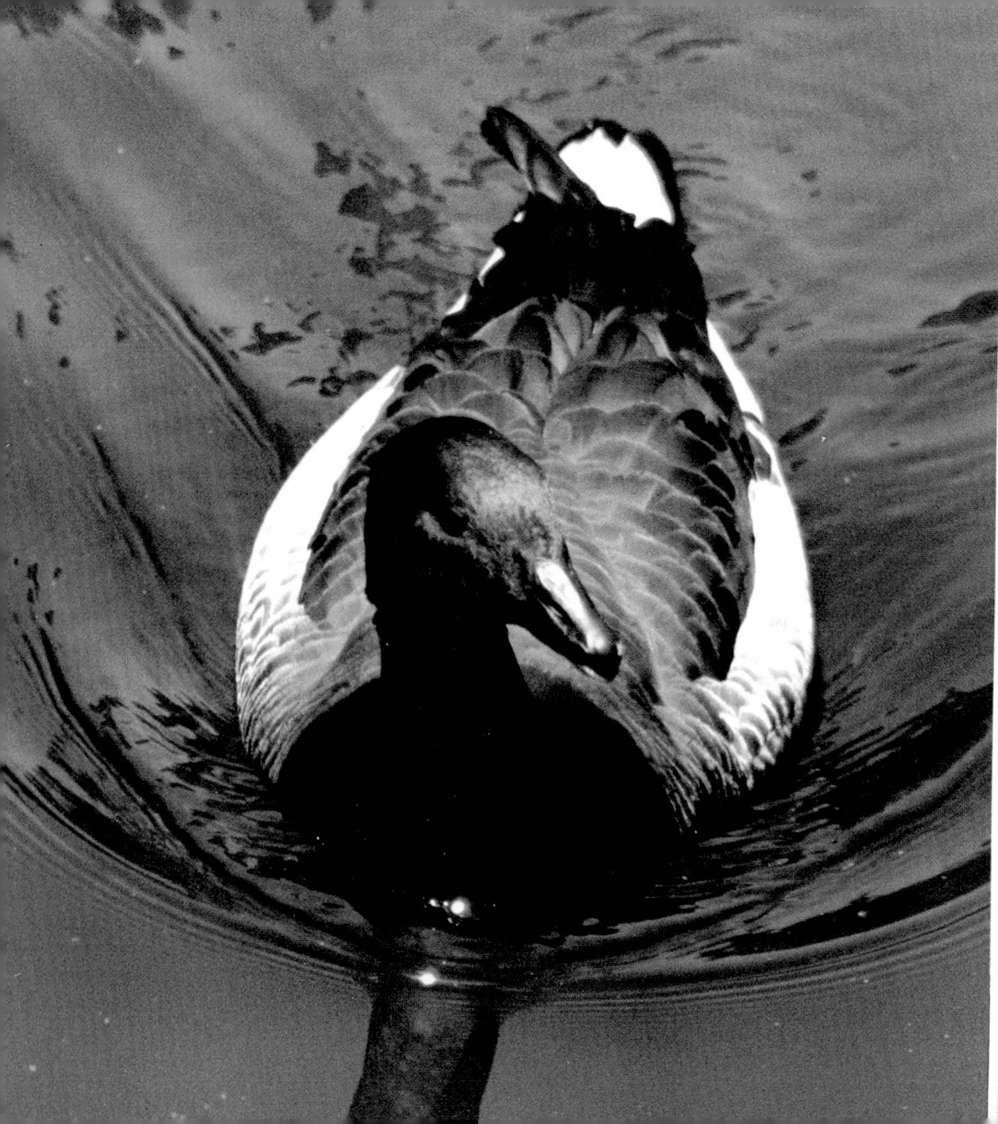
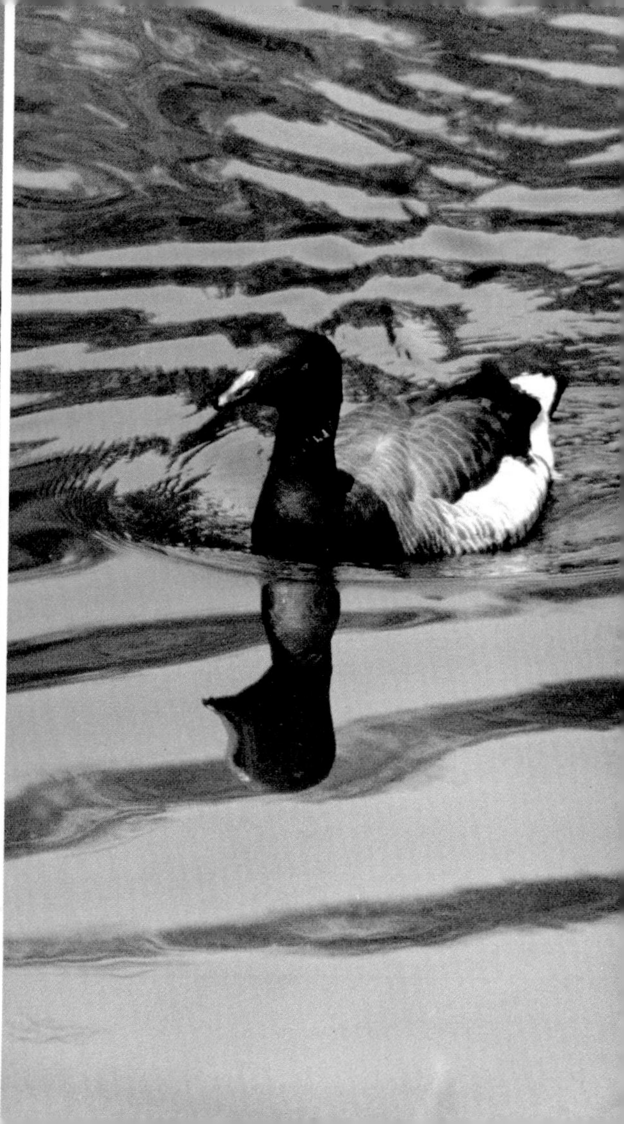

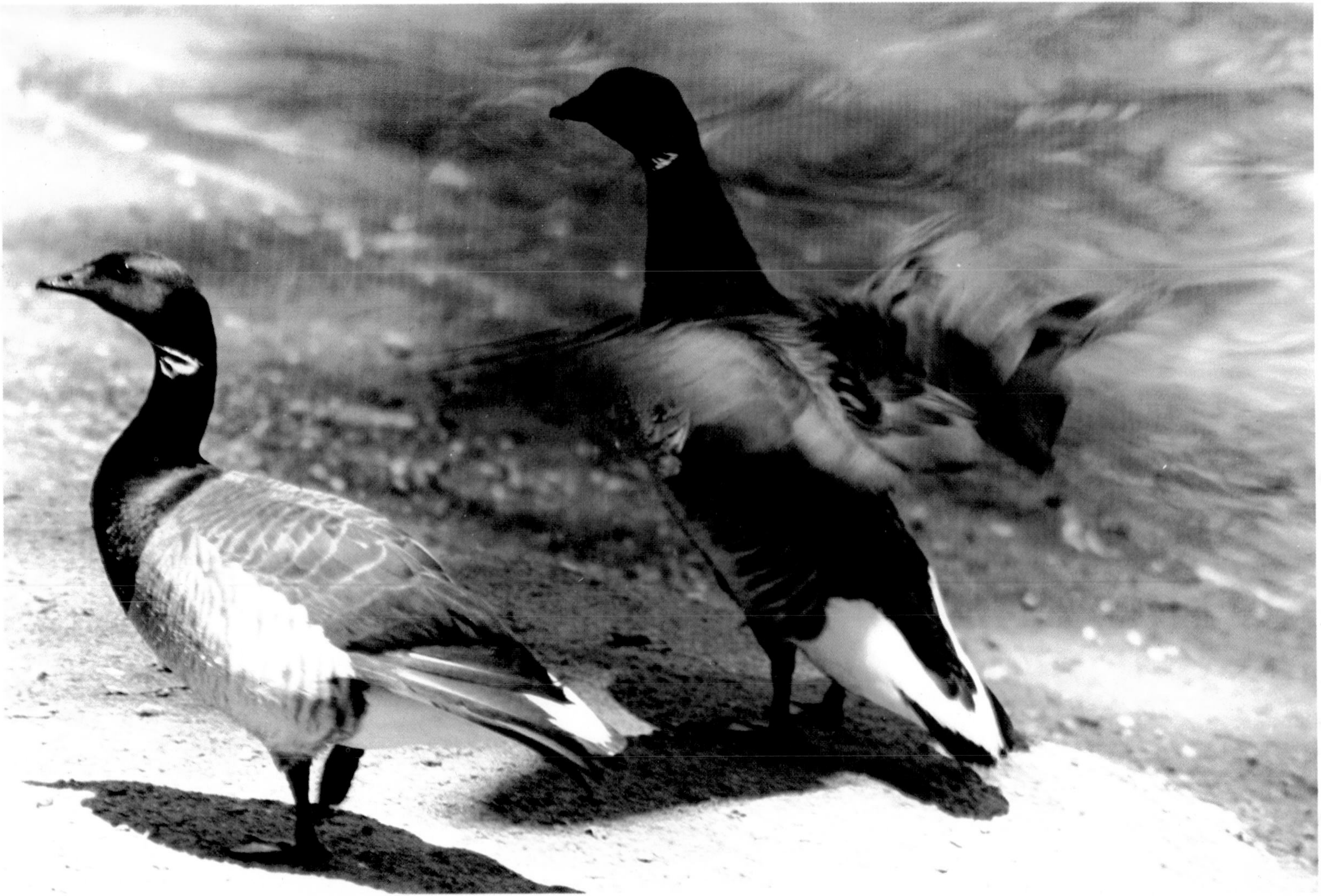

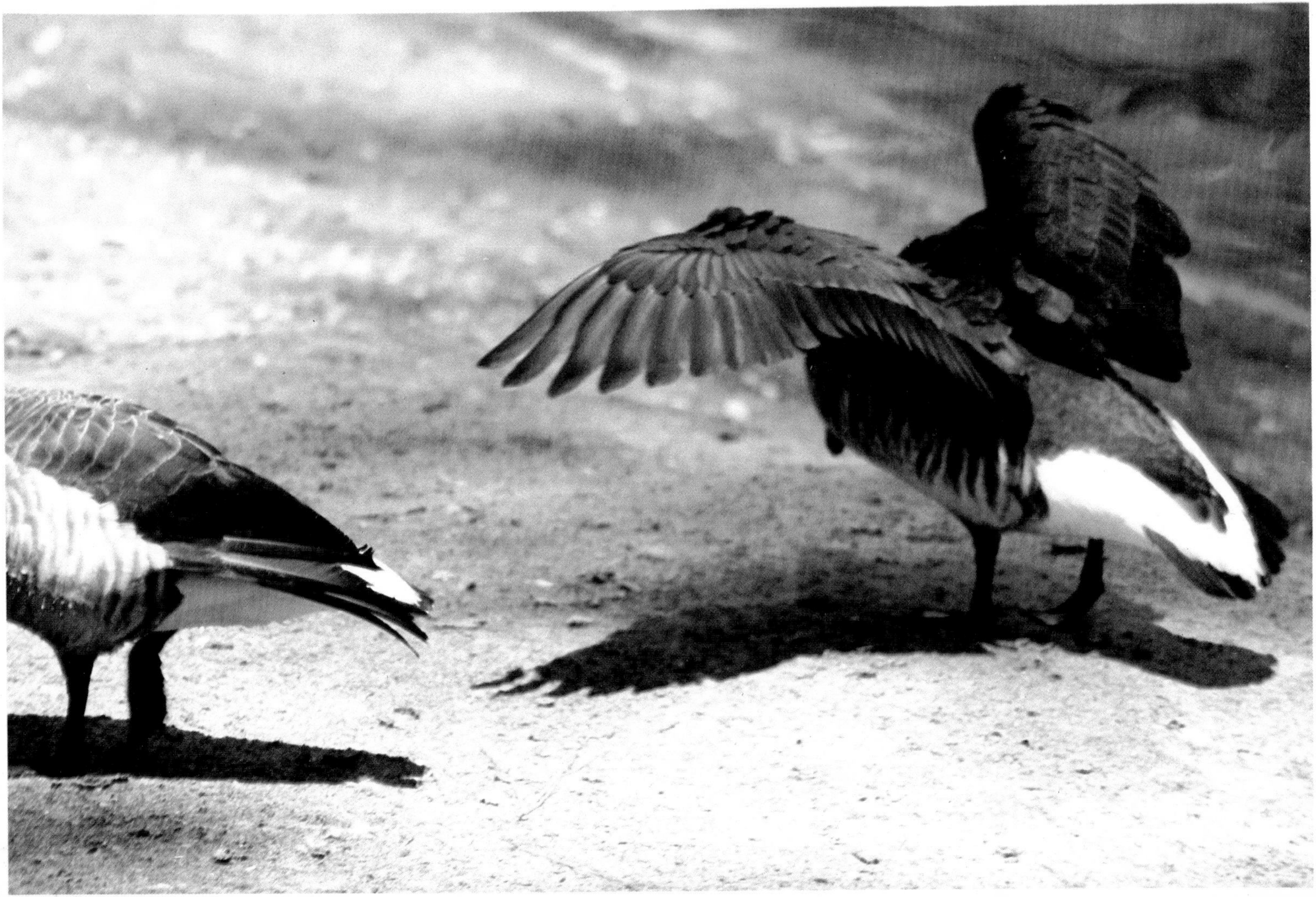

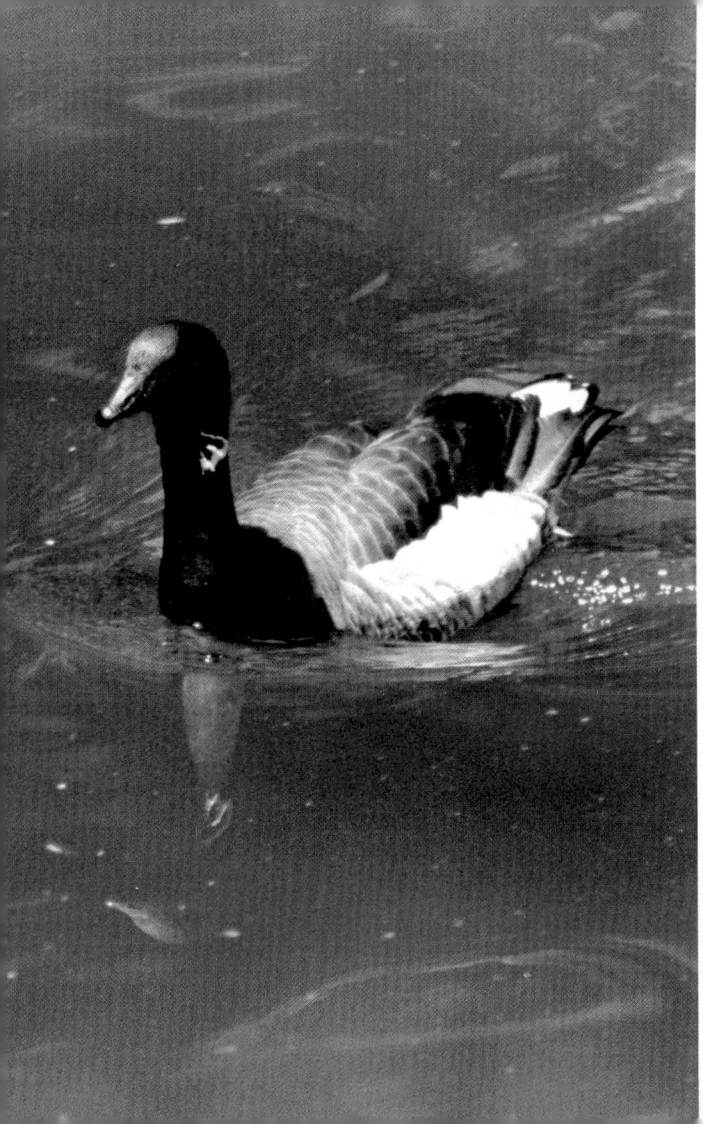

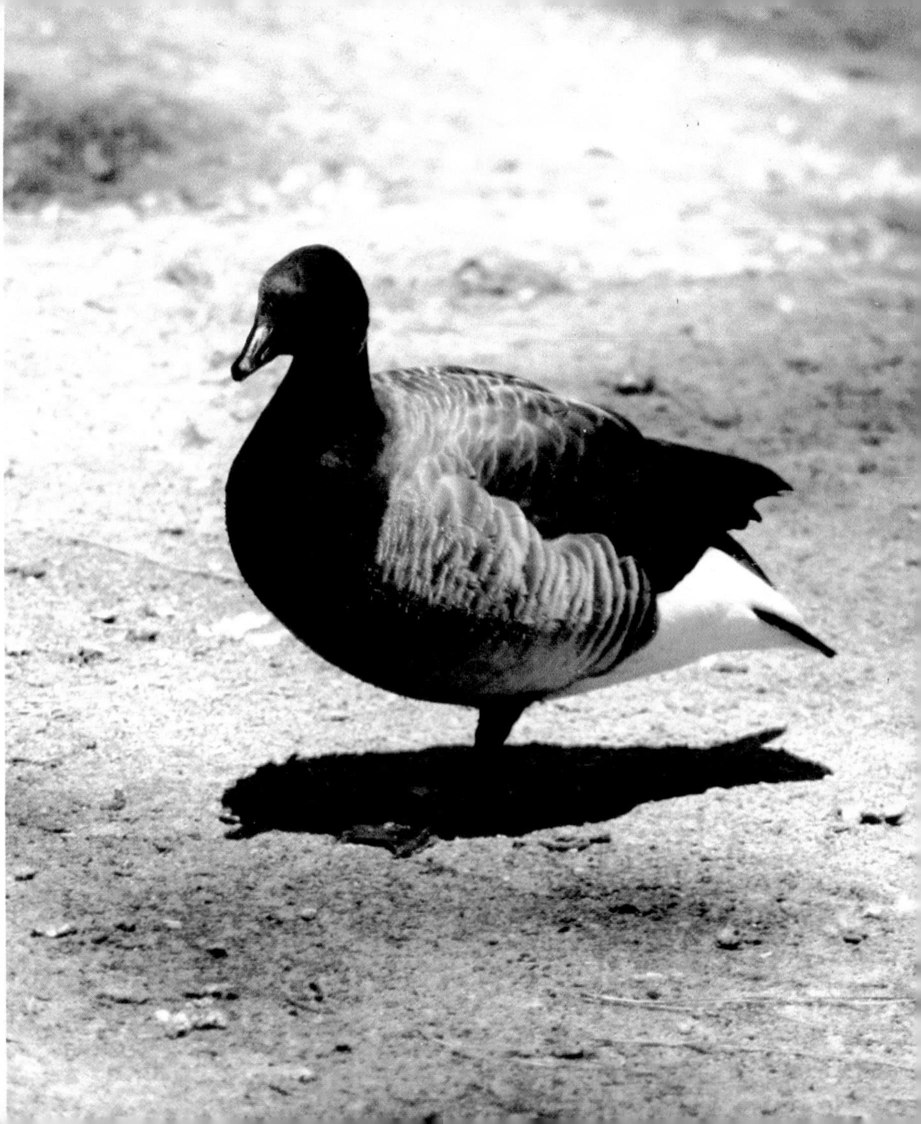

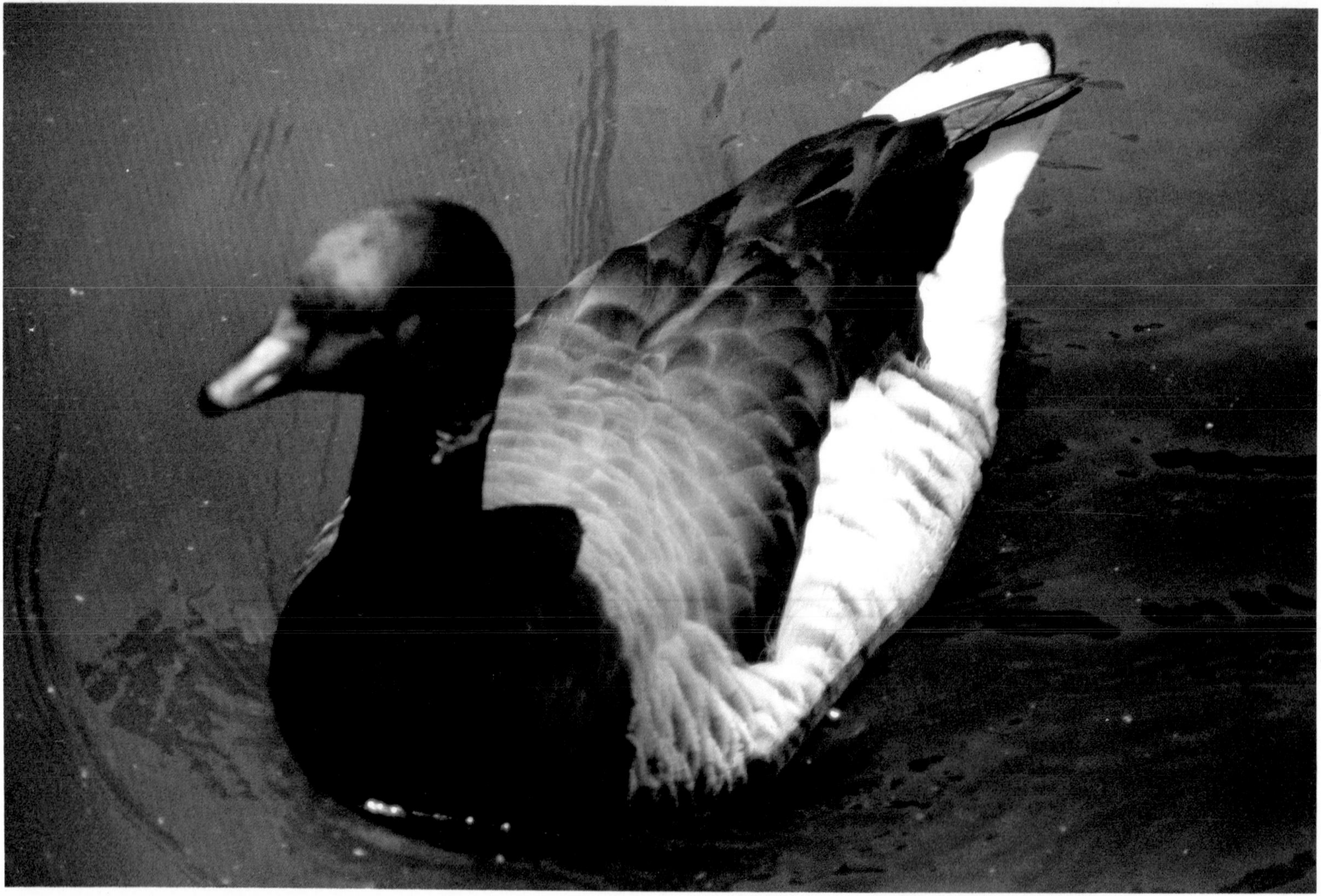

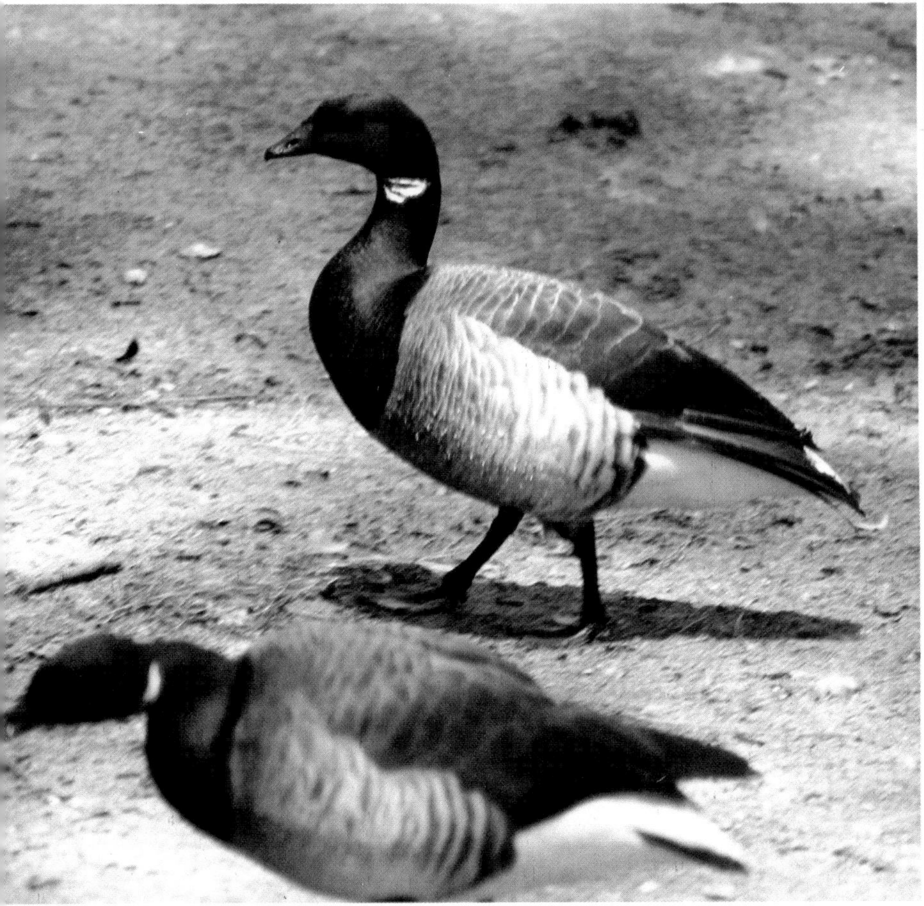
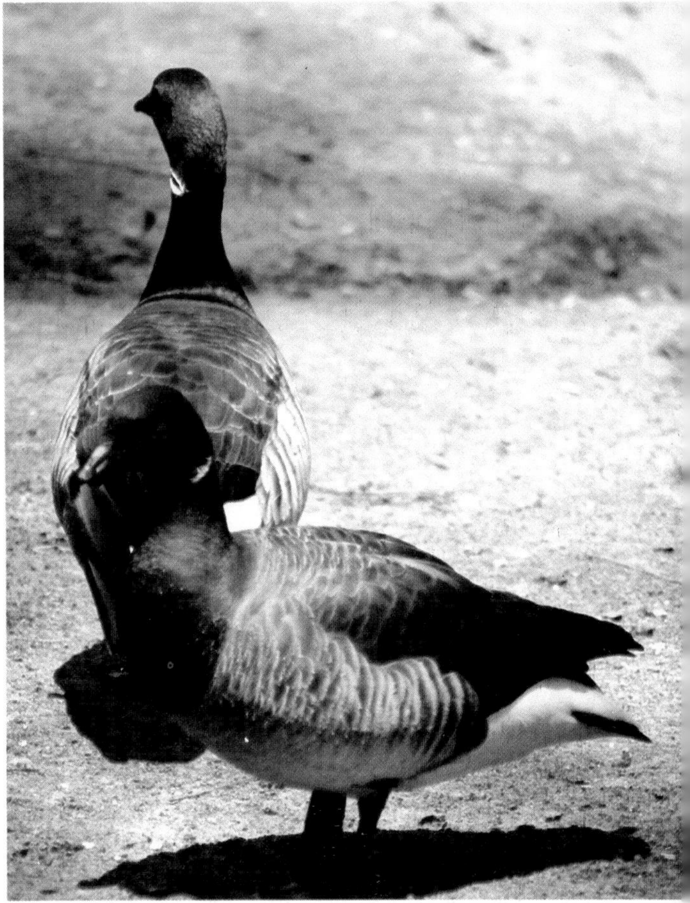

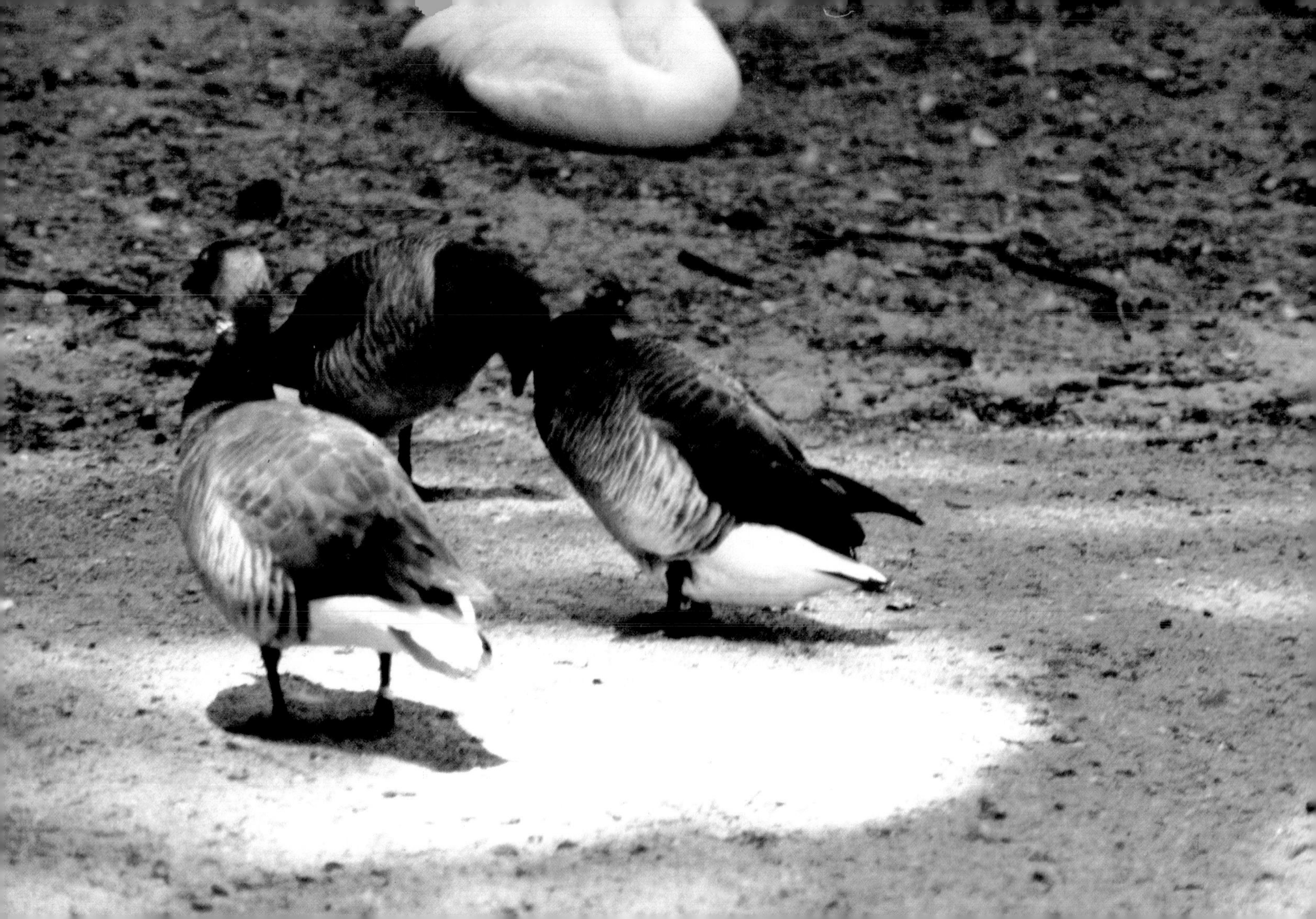

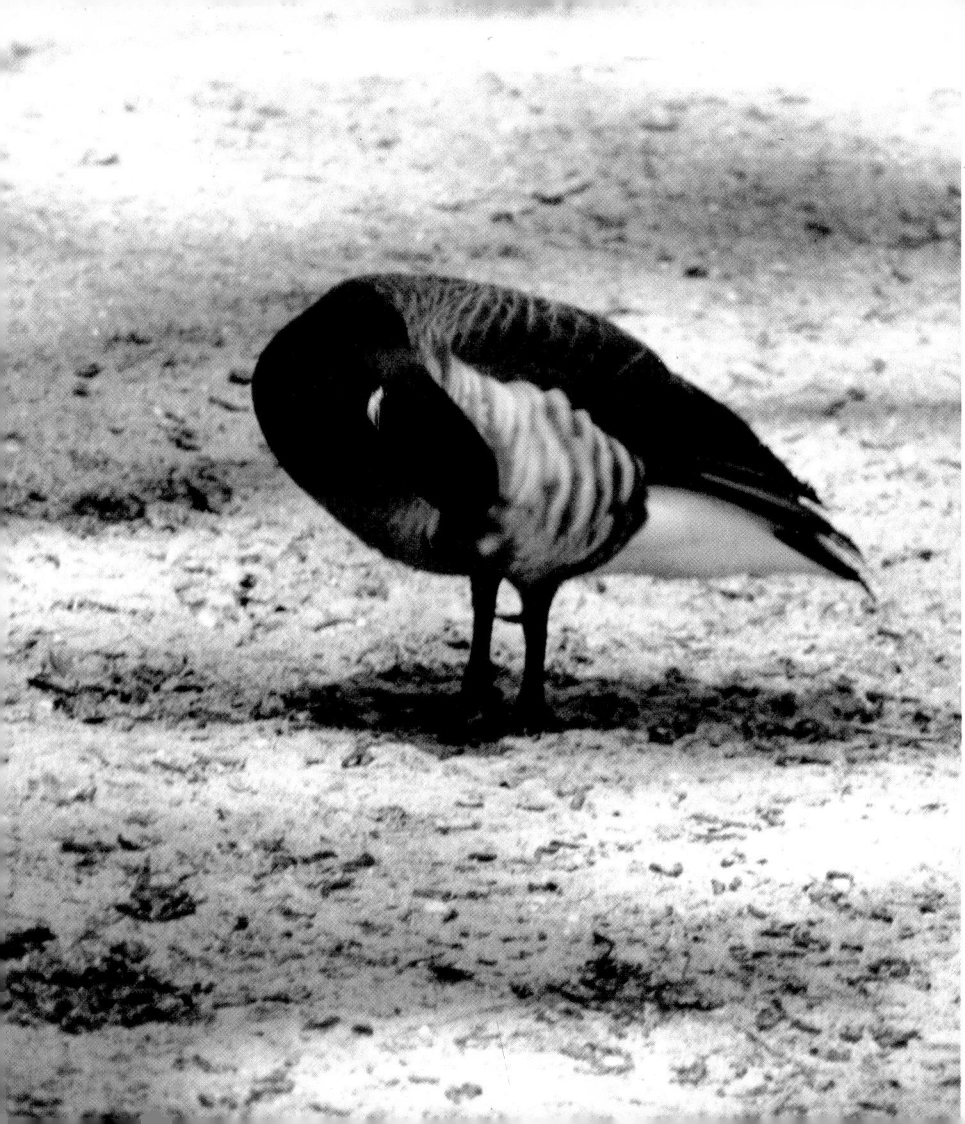
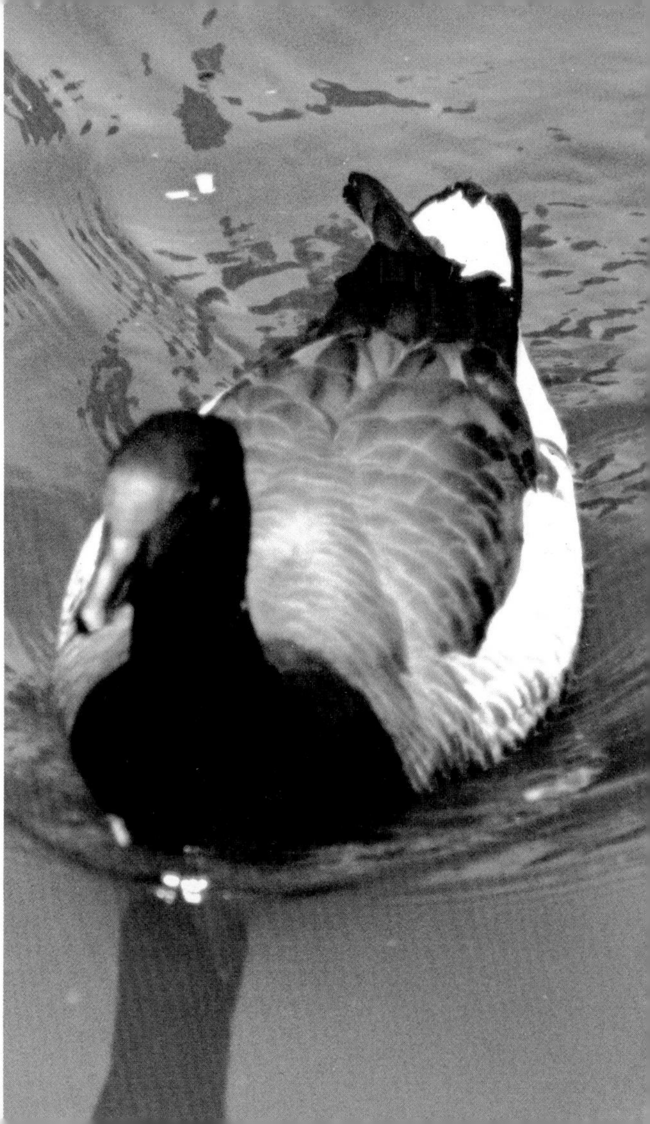

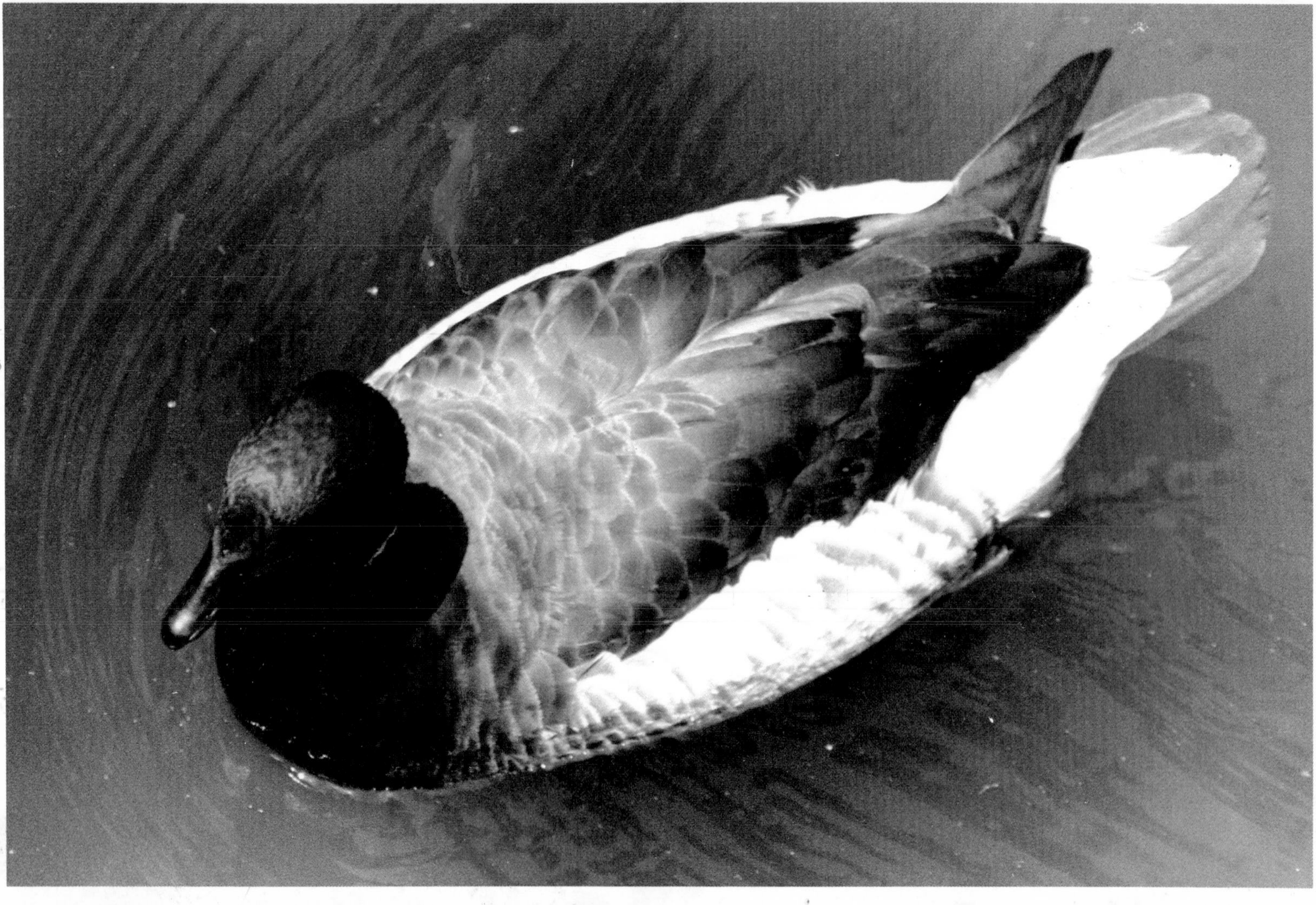